Laboratory Activities Masters for

ENVIRONMENTAL SCIENCE:
A Framework for Decision Making

HIGH SCHOOL EDITION

Daniel D. Chiras

University of Colorado, Denver

Addison-Wesley Publishing Company
Menlo Park, California • Reading, Massachusetts • New York
Don Mills, Ontario • Wokingham, England • Amsterdam • Bonn
Sydney • Singapore • Tokyo • Madrid • Bogotá • Santiago • San Juan

To the Teacher

The laboratory activities masters in this publication duplicate the research labs and activities beginning on page 681 in the student text *Environmental Science: A Framework for Decision Making.* While all of the lessons in the student text require that students write their answers on separate sheets of paper, the blackline masters in this publication provide spaces for student responses when appropriate. Nonreproducible blue answers appear in these spaces or in the margins. Teacher notes also appear in margins in nonreproducible blue.

On page 3, the table of Contents shows the student text page number of each research lab and activity as well as the page number of each corresponding blackline master in this publication.

The Learning Objectives and Means of Achieving Them, beginning on page 5, are designed to assist teachers by summarizing the essence of each laboratory lesson.

The Correlation of Research Labs and Activities with Text Chapters, on page 11, shows which laboratory lessons support each text chapter.

Activities 26 and 27 were adapted from "Science and the Scientific Method" and "Population Growth," *Addison-Wesley Biology: A Systems Approach Laboratory Manual,* by George Zahrobsky, © 1988 by Addison-Wesley Publishing Company, Inc.

ISBN 0-201-22860-2

BCDEFGHIJKL-ML-892109

Contents

Learning Objectives and Means of Achieving Them

Research Lab 1 Growing Plants

Objective/Purpose: To study the effects of the best growing conditions on a particular plant

Achieving the objective: Students choose a kind of crop and determine the best growing conditions for that particular plant.

Research Lab 2 The Effect of Salt on the Growth of Bacteria

Objective/Purpose: To study the effect of salt water on bacteria and to discover the significance of this information

Achieving the objective: Students "make" sea water and, with use of methylene blue dye, observe the interaction of bacteria and salt.

Research Lab 3 The Effect of Pollution on Seeds

Objective/Purpose: To study the effects of pollution caused by detergents on the germination of pinto bean seeds

Achieving the objective: Students soak beans for germination, then introduce different concentrations of detergents to the beans to observe the detergent's effect on seed germination.

Activity 1 Ecology Field Trip

Objective/Purpose: To provide an opportunity to review firsthand some of the important concepts of ecology and to sharpen library research skills

Achieving the objective: Class members take a short hike in a neighboring field, forest, or park, and make notes on the various ecological concepts that are observed.

Activity 2 Family Size and Population Growth

Objective/Purpose: To observe the effect of family size on population growth in the U.S. by comparing growth rates based on 2- and 3-child families

Achieving the objective: Students estimate the effects of the 3-child family (compared to the 2-child family) on the demand for natural and civil resources.

Activity 3 The Roots of Hunger

Objective/Purpose: To research the 8 contributing factors to world hunger and how each can possibly be overcome

Achieving the objective: Students break into 8 groups; each group selects 1 of the 8 factors to research. Each of the 8 groups, after research is completed, presents their findings to the class.

Activity 4 Climatograms

Objective/Purpose: To relate climate to the plant and animal communities of different areas by plotting climatograms of biomes

Achieving the objective: Students plot climatograms from given data and identify and compare the various biomes.

Activity 5 Water Research

Objective/Purpose: To promote a better understanding of local water supplies and ways to meet increasing water demands

Achieving the objective: Students meet with their local water departments to ask selected questions about their community's current water system.

Activity 6 Water Quality

Objective/Purpose: To learn to perform water-quality tests and to interpret the results of these tests

Achieving the objective: Students use a water-quality test kit to analyze water from natural and treated sources.

Activity 7 Electrical Energy in the Home

Objective/Purpose: To determine how much electrical energy is used at home, and to devise and test a plan for reducing our electrical energy use

Achieving the objective: Students calculate the number of kilowatt hours of electricity used in their homes in a 1-month period by averaging electric bills over the past 6 months. With these figures, students work toward the goal of reducing electricity consumption by 15%.

Activity 8 Recycling Waste Materials

Objective/Purpose: To identify the variety of items that we throw away that can be reused

Achieving the objective: Students collect items from home and discuss what could be made out of each type of disposable item.

Activity 9 Solar Water Heater

Objective/Purpose: To demonstrate how water can be heated and circulated by solar power using a model system

Achieving the objective: Students dye water to follow its path through a model system, then analyze why the system works and why it can be improved.

Activity 10 Solar Distillation

Objective/Purpose: To demonstrate the use of solar power to distill fresh water from salt water in a model system

Achieving the objective: Students use a solar greenhouse to observe the process of solar distillation.

Activity 11 Light Production

Objective/Purpose: To compare the amounts of light produced by fluorescent and incandescent lamps of the same wattage

Achieving the objective: Students use a light meter to measure the amount of light given off at distances of 1, 2 and 3 meters by the 40-watt fluorescent lamp and the incandescent lamp.

Activity 12 The Great Joule Robbery

Objective/Purpose: To demonstrate energy transformations and losses by building and operating a simple steam engine

Achieving the objective: Students observe the steam engine and from its operation gather information about the engine itself and gain a perspective with which to observe other common machines' transformation of energy.

Activity 13 The Greenhouse Effect

Objective/Purpose: To demonstrate the greenhouse effect and suggest ways your observations can help you design solar collectors

Achieving the objective: Students use thermometers and white and black paper to collect data on light-absorbing materials.

Activity 14 Some Properties of Soils

Objective/Purpose: To analyze and classify soil types, and to make some predictions about the behavior of soils

Achieving the objective: Students should collect soil samples from known locations and relate them to the plant and animal populations in those areas.

Activity 15 Learning About Energy Quality

Objective/Purpose: To determine whether wood or paraffin has more energy per unit weight

Achieving the objective: Students record water temperature before and after igniting both wood and paraffin separately, and then use a given formula to calculate the number of calories that are contained in 30 grams of each material.

Activity 16 Air Pollution

Objective/Purpose: To measure the particles in the air, and to determine whether certain locations have more particles than other locations and (in Part 2) to relate types of pollutants to the types of energy sources utilized

Achieving the objective: Students use petri dishes to gather particles from the air, and use data tables to record their results. They also research local air polluters and propose plans for reducing or eliminating 1 type of air pollution.

Activity 17 Wet-Scrubber

Objective/Purpose: To become familiar with a wet-scrubber by building and using one

Achieving the objective: Students construct a wet-scrubber model and observe how it functions as a pollution control.

Activity 18 Temperature Inversions and Air Pollution

Objective/Purpose: To model temperature inversions and (in Part 2) to demonstrate how temperature inversions trap air pollutants

Achieving the objective: Students construct an apparatus to observe how a temperature inversion affects air flow patterns and traps pollutants.

Activity 19 Nuclear Energy Viewpoints

Objective/Purpose: To consider the pros and cons of the use of nuclear power

Achieving the objective: Students research nuclear energy, then invite several qualified speakers to discuss why they favor/oppose nuclear energy.

Activity 20 Ecological Debate

Objective/Purpose: To consider how humans fit into the ecological world

Achieving the objective: 2 5-member debate teams take opposing sides of how humans fit into the ecological world, with the rest of the class taking notes on the questions posed and the responses of the 2 teams. Class members also make personal comments concerning the debate.

Activity 21 Popularizing Environmentally Favorable Lifestyles

Objective/Purpose: To devise ways to make environmentally favorable lifestyles more popular

Achieving the objective: Students focus on the positive aspects of a conserving lifestyle and present their findings to the class.

Activity 22 Biodegradability of Solids in a Mini-Landfill

Objective/Purpose: 1) To determine the rates of decomposition of various substances in a model landfill, and 2) to consider the effectiveness of landfills for waste disposal

Achieving the objective: Students set up a mini-landfill in a glass with potting soil and various substances to observe the biodegradability of selected solids over a 4-week period.

Activity 23 Does Insulation Really Help to Conserve Heat?

Objective/Purpose: To determine how well different types of insulation reduce heat loss in a model system

Achieving the objective: Students use insulated and uninsulated beakers with thermometers to gauge comparative temperatures.

Activity 24 Thermal Pollution

Objective/Purpose: To determine the effects of thermal pollution on goldfish and other fish

Achieving the objective: Students vary water temperature to test the effects of these changes on goldfish.

Activity 25 Investigating the Biological Community of a City

Objective/Purpose: To study a city as a biological community by identifying the plants and animals of the city, determining their relationships to each other and to human beings, and tracing the energy flow of the city as a whole

Achieving the objective: Students use binoculars and field guides to study their neighborhood's biological community.

Activity 26 Science and the Scientific Method

Objective/Purpose: To practice gathering scientific information, testing ideas, and solving problems

Achieving the objective: Students read a fictional experiment, determine the operating hypothesis, assess the procedure used and data gathered, and form conclusions.

Activity 27 Understanding Population Growth

Objective/Purpose: To learn how the world's population grows and to explore how this growth affects the environment

Achieving the objective: Students graph the projected population of the world and apply mathematical concepts using the graphed data.

Activity 28 Land Use

Objective/Purpose: To identify school and community open spaces and consider the consequences of developing them

Achieving the objective: Students list open areas on campus and in the community, and describe the benefits and the drawbacks of developing these sites.

Activity 29 The Effects of Toxics on Living Organisms

Objective/Purpose: To investigate the effect of a toxic substance on an aquatic organism

Achieving the objective: Students create 8 petri dish environments containing *Daphnia,* then introduce various concentrations of copper sulfate and observe their effects on the aquatic organisms.

Activity 30 Food Additives

Objective/Purpose: To identify food additives and their functions

Achieving the objective: Students identify different additives in various foods and assess their advantages and disadvantages.

Activity 31 Government and the Environment

Objective/Purpose: For students to identify environmental issues of interest to them and to communicate their interests to appropriate public officials

Achieving the objective: Students outline environmental issues of concern to them, then write letters expressing their ideas to public officials.

Correlation of Research Labs and Activities with Text Chapters

Each of the research labs and activities contained in this section explores a current environmental problem while providing experience with one or more scientific processes. The research labs, in particular, give specific practice setting up controlled experiments, making independent decisions regarding procedures, and hypothesizing. Many of the activities, as well as two of the research labs, are related to several text chapters, as indicated on the chart below.

When more than one exercise supports a text chapter, the **bold** number indicates the primary, or core, laboratory experience for that chapter.

	Text Chapter	Research Lab	Activity
1	Environmental Science: Reshaping Society		26
2	Framework for Understanding Environmental Problems		**5,** 20, 28
3	Principles of Ecology: Ecosystem Structure		1, **4,** 20, 25
4	Principles of Ecology: Ecosystem Function		**1,** 20, 25
5	Principles of Ecology: Ecosystem Balance and Imbalance		25
6	Population Growth and Its Impact		2, **27**
7	Limited Population Growth		**2,** 3, 27
8	Problems Facing World Agriculture	**1**	3, 14
9	Feeding the World's People	1	**3**
10	Plants and Wildlife: Imperiled Resources	**3**	13
11	Wilderness, Parks, and Other Nonurban Land Uses		28
12	Water Resources	**2**	5, 6
13	Today's Energy: Fossil Fuels		**15,** 23
14	Future Energy: Nonrenewable Resources		7, 11, 12, **19**
15	Future Energy: Renewable Resources		**9,** 10, 13, 22
16	Mineral Resources		**6,** 8, 22
17	Environmental Decision Making: Risk Assessment and Ethical Analysis		19, **20,** 21, 28, 31
18	Toxic Substances and Environmental Health		29
19	Air Pollution		13, **16,** 18
20	Water Pollution	3	6, **17,** 24
21	Personal and Occupational Pollution		30
22	Hazardous Wastes		8, **22**
23	Environmental Ethics: The Foundation of a Sustainable Society		20, **21,** 28
24	Economics and the Environment		3, 5, **7**
25	Politics, Government, and the Environment: Toward a Sustainable Society		31

Safety in the Classroom and Field

Safety guidelines in the laboratory

1. Wear a laboratory apron to protect your clothing. Wear safety glasses or goggles when handling chemicals or heating substances with the laboratory burner.

2. Always conduct your experiments with someone else present and under adult supervision.

3. Be familiar with the location of safety equipment. Some important pieces of safety equipment are:

a) First aid kit
b) Fire extinguisher
c) Emergency shower
d) Fire blanket
e) Eye-wash station

4. Notify your teacher *immediately* if any chemicals, such as concentrated acid or base, are spilled. Also immediately notify your teacher of any accidents in the laboratory.

5. Never eat or drink while doing a laboratory experiment.

6. Wash your hands with soap and water after handling chemicals, cultures, plants, or animals. Keep your hands away from your face and your eyes.

7. Tie back long hair and roll up long sleeves when working with flames or chemicals.

8. Be sure that any electrical cords are not in a place where someone can trip over them.

9. Be sure that all laboratory surfaces, equipment and tools are cleaned after you are finished with them.

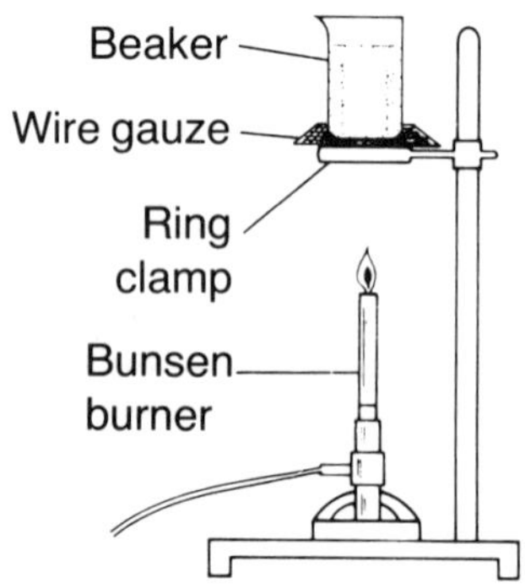

10. Turn off hot plates and laboratory burners when they are not being used.

11. When heating, only use glassware that is meant to be used for heating.

12. When heating flasks or beakers over the laboratory burner, use a ringstand set-up with a square of wire gauze.

13. Use a water bath to heat solids.

14. When heating a substance, always point the test tube away from you and any other person.

15. When heating with a laboratory burner, gently move the test tube over the hottest part of the flame.

16. Use tongs or hot pads when handling hot equipment. Use a test-tube holder when heating substances in a test tube.

17. Do not pour hot liquids into plastic containers.

18. Do not use chipped or broken glassware. Discard broken or chipped glassware according to your teacher's instructions.

19. Let glassware air-dry rather than drying any glassware with a towel.

20. Do not use direct sunlight as a light source for a microscope.

Safety guidelines for field trips

1. Your own common sense will be your best guide to dressing for field work. For example, sturdy walking shoes with good tread on the soles are safer to wear than open sandals. Clothes should also protect you from the cold.

2. Ask people who know the area you are planning to visit about the possibility of running into troublesome animals, such as ticks, or poisonous snakes and spiders. Learn the locations of forest ranger stations and first aid units. Study a map of the area and note where streams, rivers, and lakes are located.

3. If poisonous snakes are present, stay on trails or in clearings. Sudden confrontations which surprise the snake are the cause of most snakebites. Snakes and other wild animals will avoid you whenever they can.

4. Watch out for poisonous plants such as poison ivy, poison oak, or poison sumach. Avoid any contact with these plants. Even though you may think you are not susceptible to poisoning from these plants, you may find that you are at certain times.

Poison oak

5. Watch out for deep water, swift currents, or quicksand. Never go into the water alone. Carry life vests. Wear rubber boots when necessary.

6. Never look directly at the sun, and never use optical instruments such as telescopes or field glasses to look at the sun. Serious damage to the eyes can result.

7. Never hike alone into the wilderness. Stay with large groups and follow established trails. Carry a compass and a supply of water and food.

8. Carry a first aid kit for treatment of cuts, bites, bruises, and fractures.

9. If you puncture your skin, get a tetanus inoculation as soon as possible.

10. Do not drink water unless you are sure it is free of dangerous chemicals and organisms.

Common Equipment in the Laboratory

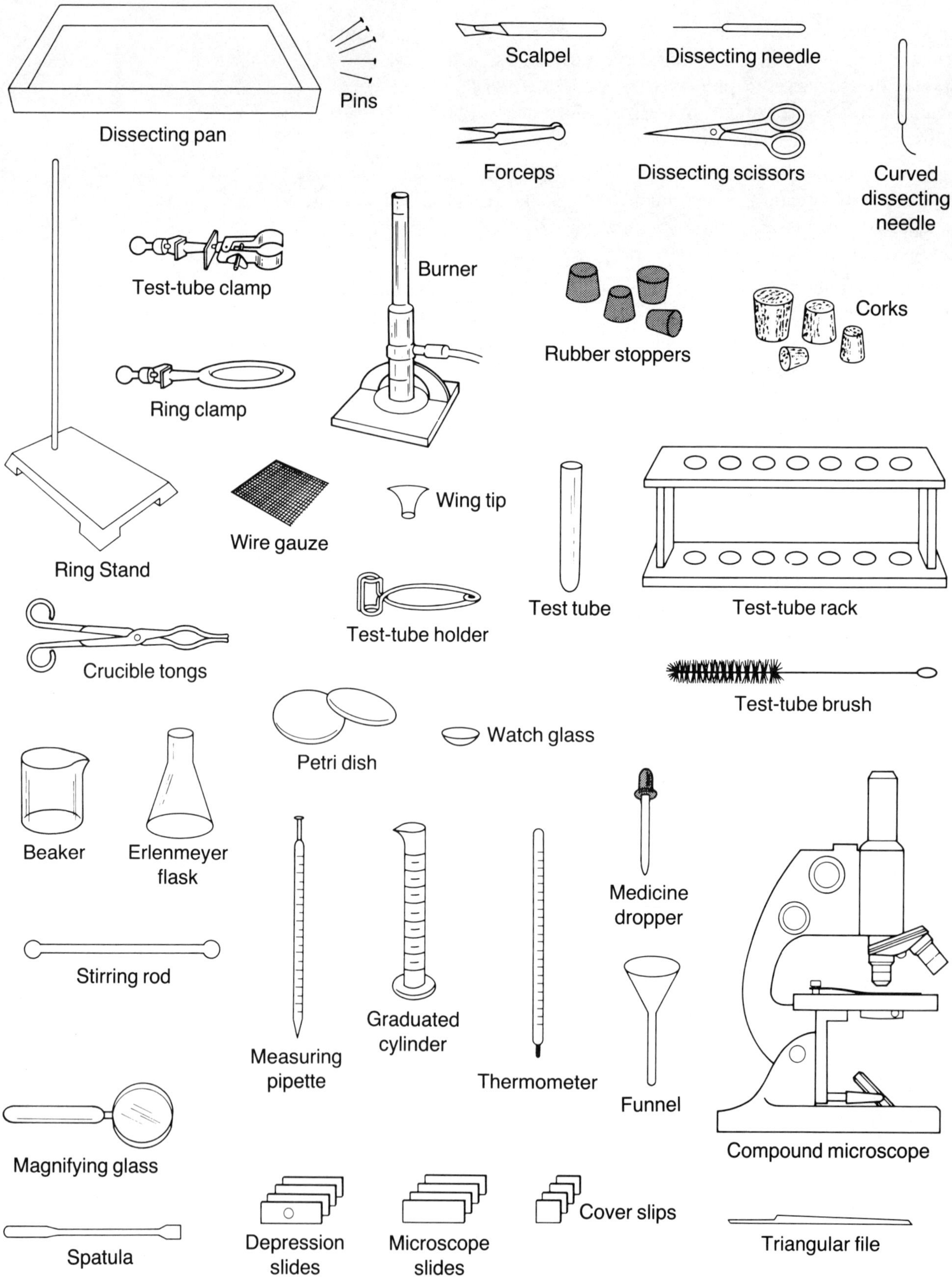

Common Equipment in the Field

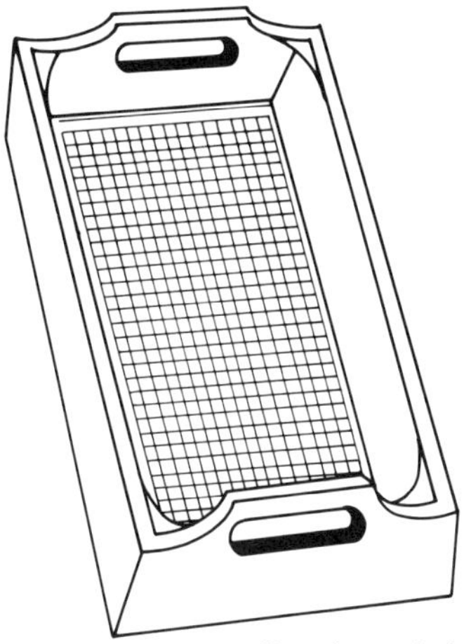

Sand-mud sieve

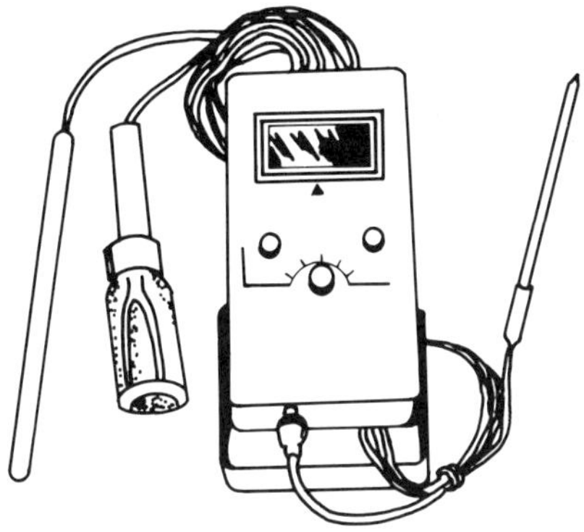

Digital pH/temperature/conductivity meter

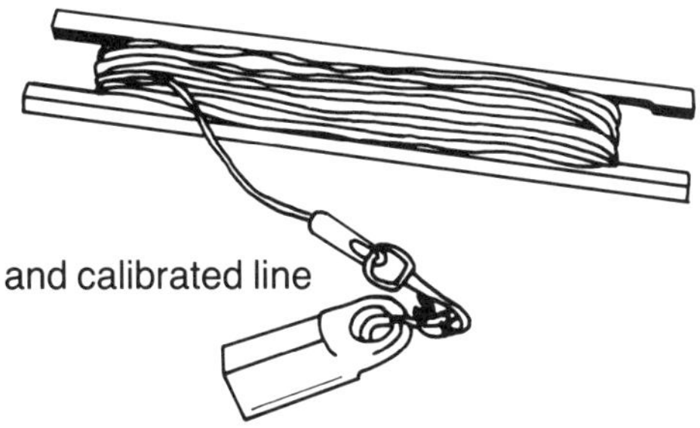

Sounding lead and calibrated line

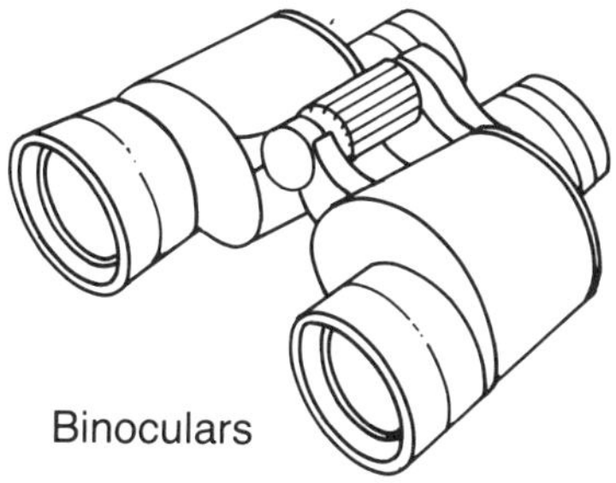

Binoculars

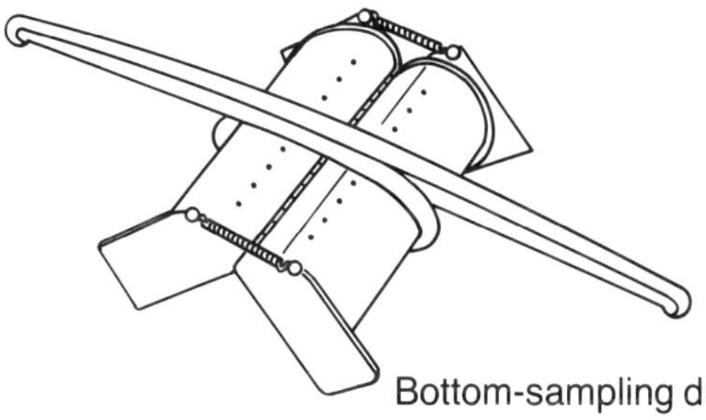

Bottom-sampling dredge

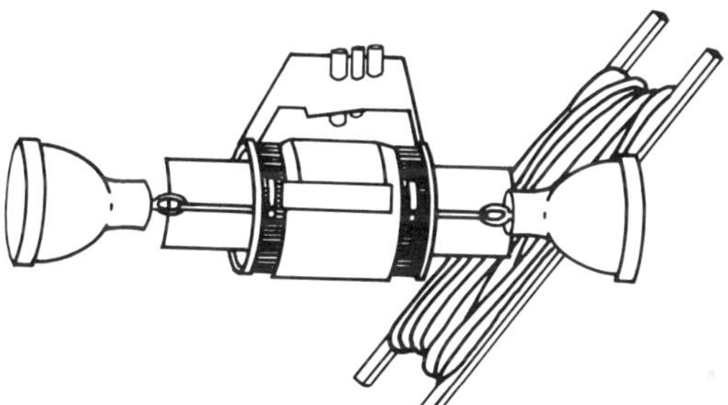

Water-sampling bottle

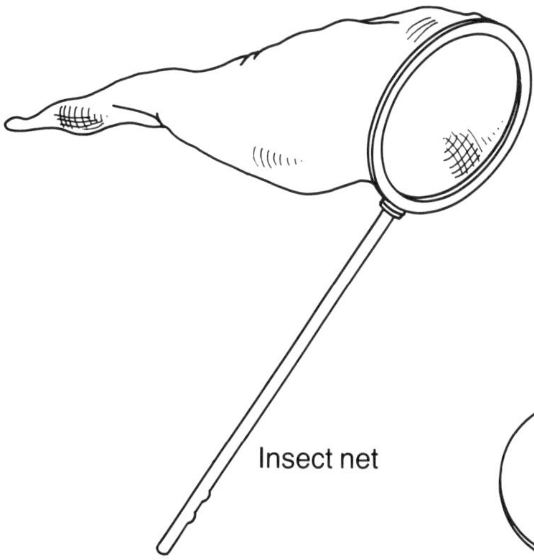

Insect net

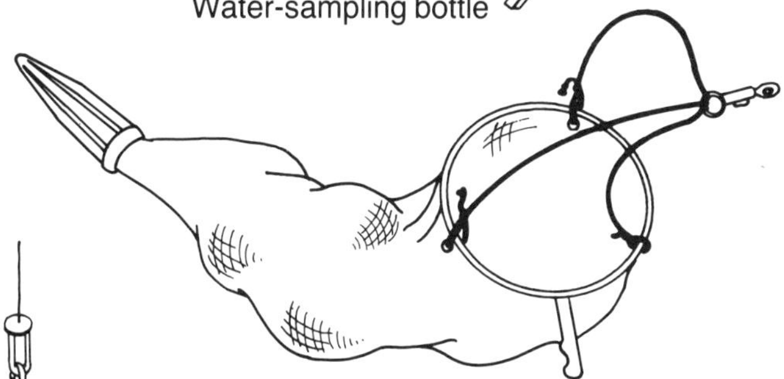

Plankton net

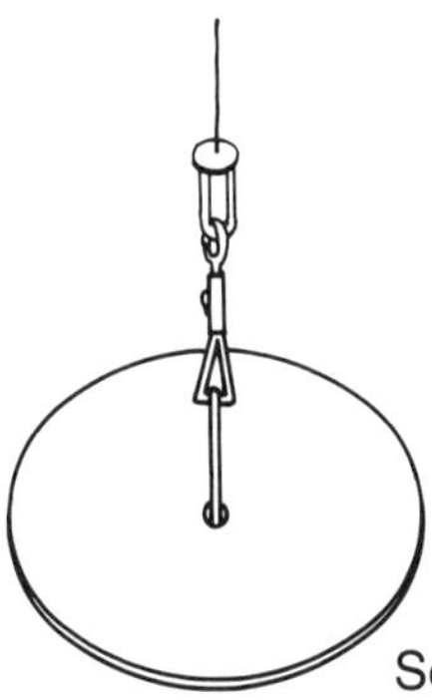

Secchi disk

Safe Use of Equipment

Using heating equipment safely

1. Wear special gloves or use tongs, pot holders, or hot pads to handle hot objects, materials, and containers.

2. Before using a flame or hot plate, roll up loose sleeves, or band them lightly. Also, tie back long hair that might touch the heating source.

3. Arrange to use heat only in areas near fire extinguishers, fire blankets, and first aid kits.

4. Read and follow procedures on page 19 for lighting a laboratory burner.

5. Turn burners and heaters off whenever they are not in use, even if they are going to be used again in a minute or two.

6. Heat only glassware, such as Pyrex® or Kimex®, that is made to be heated. Keep plastic labware away from heat, and use heat-resistant glassware, not plastic, to hold hot liquids.

7. Follow a water bath procedure to heat solids. Never risk an explosion by directly heating rocks or minerals.

8. Heat flasks or beakers by clamping them on a ringstand with a wire gauze platform between the glass and the flame.

9. Never reach for something so that your arm or body is over the open flame of a laboratory burner.

10. Allow plenty of time for hot equipment and materials to cool before touching them. Heat-resistant glassware can be slowly cooled by first rinsing with water of slightly lower temperature than the glassware, followed by rinsing with lower and lower water temperatures until the glassware is cooled. Other hot equipment should be placed on wire gauze mats to air cool.

11. Review the first aid procedures for burns ahead of time so that you can act quickly, calmly, and effectively in a burn emergency. Mild burns should be immediately soaked in ice-cold water.

Measuring length

In the metric system, a metric ruler is divided into centimeters. Each centimeter is divided into 10 millimeters. To measure the length of an object, place the metric ruler along the side of the object to the nearest millimeter. If the length is between two marks, use the length to the nearest millimeter mark.

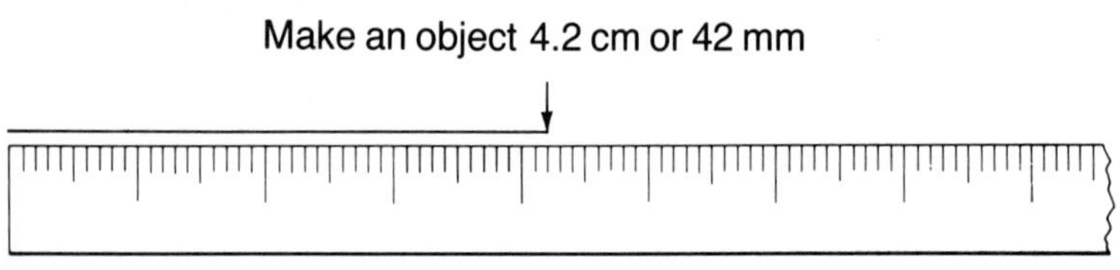

Measuring mass

A laboratory balance is used to find the mass of an object. There are several types of laboratory balances. But, each type of balance has many of the same parts. Each balance has beams on which the mass of an object will be indicated. Beam balances may have one, two, or three beams.

To measure the mass of an object, follow these instructions:

1. Place the balance on a flat, level surface.

2. Be sure to move all riders back to the zero point.

3. Check to see that the pointer is at the zero point. If the pointer is not at zero, use the adjustment knob on the balance so that the pointer is at the zero point.

4. Place the object on the pan. If you are using a double-pan balance, place the object on the left-hand pan.

Double-pan balance

Place the object on the left-hand pan. Use the right-hand pan to place the standard masses. Place the standard masses onto the right-hand pan until the balance tips to the right. Then, remove the smallest weight that you placed on the right-hand pan. Use the riders on the beam or beams to balance the two pans. The pointer will be at the zero point when the two pans are in balance. The mass of the object is equal to the total of the standard masses on the right-hand pan and the readings on the beam or beams. *Note: To find the mass of some objects, you may not have to use standard masses. Only the riders on the beam or beams may have to be used.*

Triple beam balance

When you place the object on the pan, the pointer should move above the zero point. Using the rider on the beam divided into 10 equal parts of 10 grams each, move the rider until the pointer drops below the zero point. Then, move the rider back one notch. Repeat this with the middle rider. Then, use the rider on the beam divided into tenths of a gram to bring the pointer to the zero point. The mass of the object is the total of the readings on the three beams. *Note: You may not always have to use all three beams to find the mass of an object.*

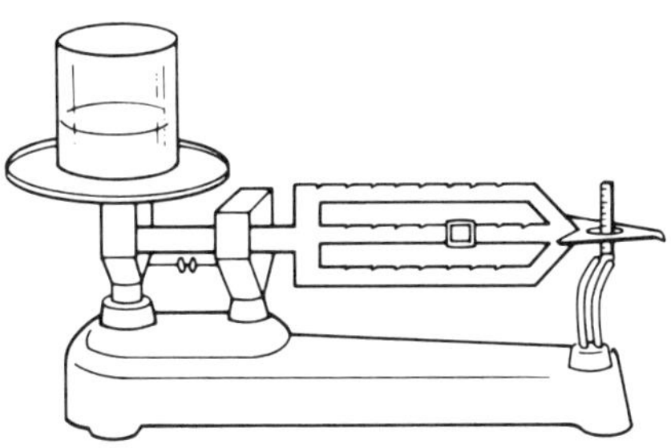

Triple beam balance

Measuring temperature

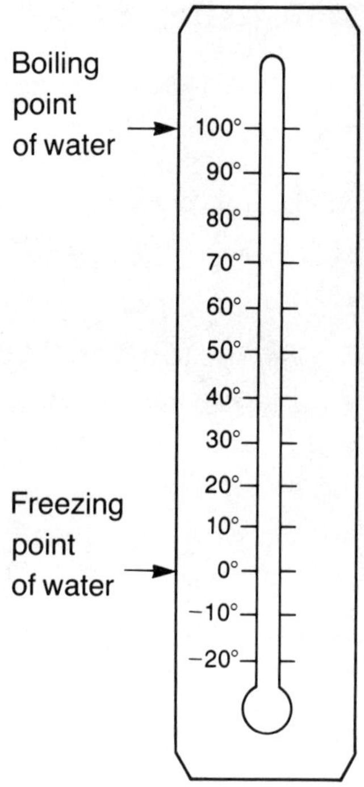

Celsius thermometer

In the metric system, the unit of measurement used for temperature is the Celsius degree. On the Celsius scale, the freezing point of water is 0°C and the boiling point of water is 100°C. Between these two points, there are 100 degrees. Each small division on a Celsius thermometer represents a degree. Divisions below 0°C are read as negative. Divisions above 0°C are read as positive. But usually the term positive is not used.

Before you measure the temperature of a liquid, be sure that the thermometer that you use covers the expected temperature range. Most thermometers that you will use in the laboratory have a temperature range of -20°C to 100°C.

Place the bulb of the thermometer into the liquid. Be sure that the thermometer does not touch the bottom or the sides of the container. Caution: *Do not stir the liquid with the thermometer.* The column of mercury in the thermometer will begin to rise or to fall. Once the mercury stops moving, read the temperature on the scale.

Measuring volume

In the laboratory, you will probably use a graduated cylinder to measure the volume of liquids. The unit of volume in the metric system is the liter. The graduated cylinder will probably be divided into units called milliliters. When you read the volume of the liquid in the graduated cylinder, place the cylinder on a flat, even surface. View the graduated cylinder from the sides at a right angle to your eyes. Be sure that you read the volume along a horizontal sight line.

A liquid in a glass graduated cylinder will have a curved surface. This curved surface is called the meniscus. If you are using a plastic container, there will not be a meniscus. Read the volume of the liquid from the bottom of the curve of the meniscus. Some liquids, such as mercury, that do not wet the glass of the cylinder form a meniscus that curves the other way. For these liquids, read the volume from the top of the curve of the meniscus.

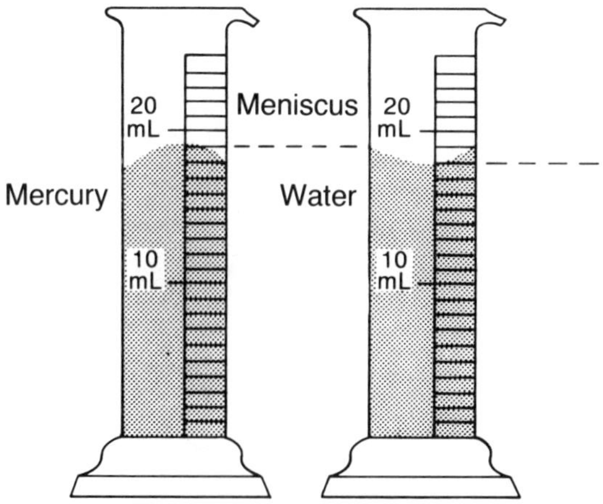

Using a laboratory burner

Connect the hose of the burner to the laboratory gas outlet. Hold a lit match along the edge of the barrel at the top of the burner. With your other hand, open the valve of the gas outlet. You may also use a striker to make a spark instead of using a lit match.

The flame size can be adjusted by turning the gas-supply valve on the burner or by partly closing the gas-outlet valve. The gas-air mixture in the barrel can also be adjusted by turning the air-adjustment sleeve. If too much air is present, the flame may go out. If too little air is in the gas-air mixture, the flame will be yellow. You should have a blue flame about 3 cm to 6 cm tall.

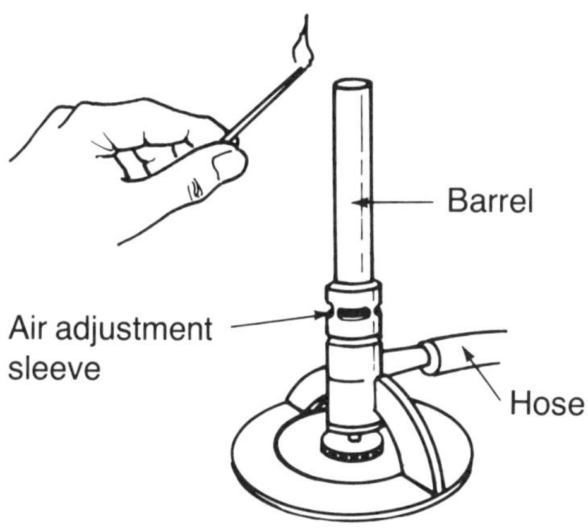

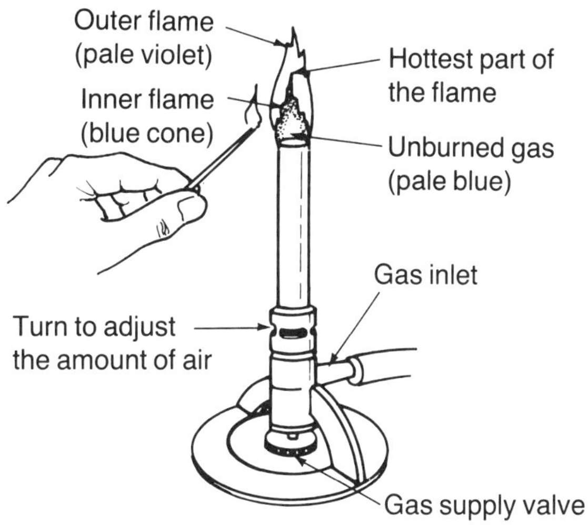

Inserting glass tubing or a thermometer into a stopper

The following procedure can be used to insert either glass tubing or thermometers into a rubber stopper. Caution: *This procedure should be performed carefully. Do not force the tubing or the thermometer.*

Obtain a rubber stopper with holes. Check the size of the holes in the stopper. The holes should be slightly smaller than the diameter of the glass tubing or thermometer.

Place a few drops of water or glycerol on the end of the glass tubing or thermometer. The water or glycerol acts as a lubricant and should make the tubing or thermometer go through the stopper hole more easily. Wrap the tubing or thermometer in a paper towel. Then, hold the stopper with another paper towel. Holding the tubing or thermometer in one hand and the stopper in the other, gently push the tubing or thermometer through the stopper. Use a gentle twisting motion as you push. Caution: *Do not hold the tubing or thermometer or the stopper so that the end of the tubing or thermometer points toward or pushes against the palm of your hand.*

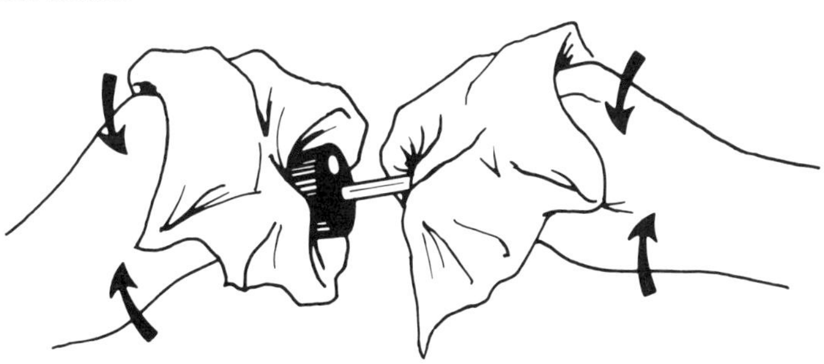

Transferring liquids

To transfer a liquid from one container to another, follow these guidelines. Place a stirring rod in the container into which you are pouring the liquid and hold the rod against the container's lip. Slowly pour the liquid down the stirring rod.

If the bottle from which you are pouring a liquid has a glass stopper, remove the stopper by grasping it with the back of your middle and index fingers (do not place the stopper on the laboratory table). Then use that same hand to grasp the bottle and pour from it.

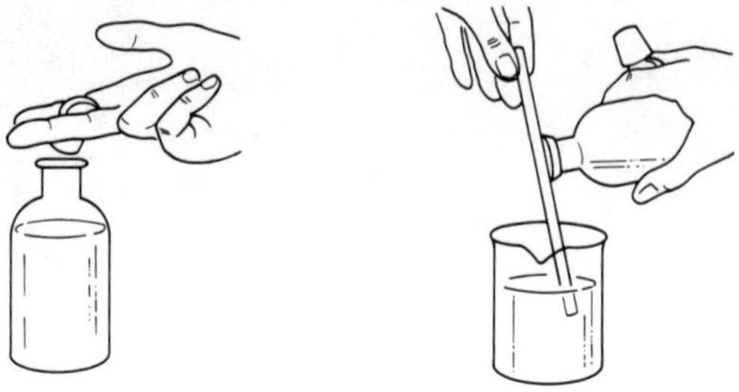

Filtering

One of the most common ways to separate a solid mixed with a liquid is filtration. The solid-liquid mixture is filtered through a funnel of filter paper. The filter paper traps any solids floating or suspended in the liquid. To prevent splashing or an overflow of the funnel, gently pour the liquid down a stirring rod.

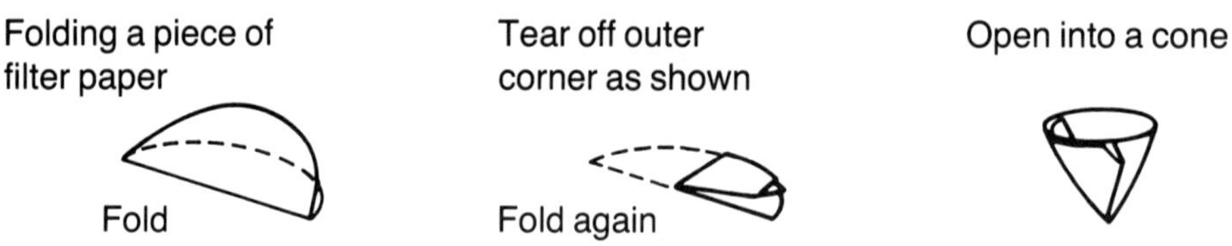

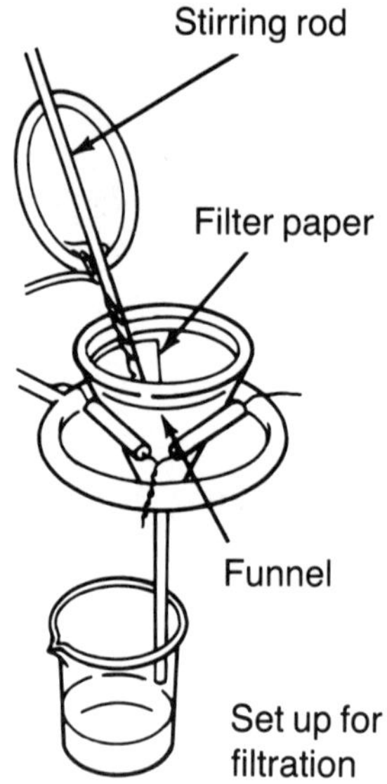

Set up for filtration

Care and use of a compound microscope

One of the most important tools that you can use to study plants and animals is the microscope. Before you begin, study and learn the parts of a compound light microscope.

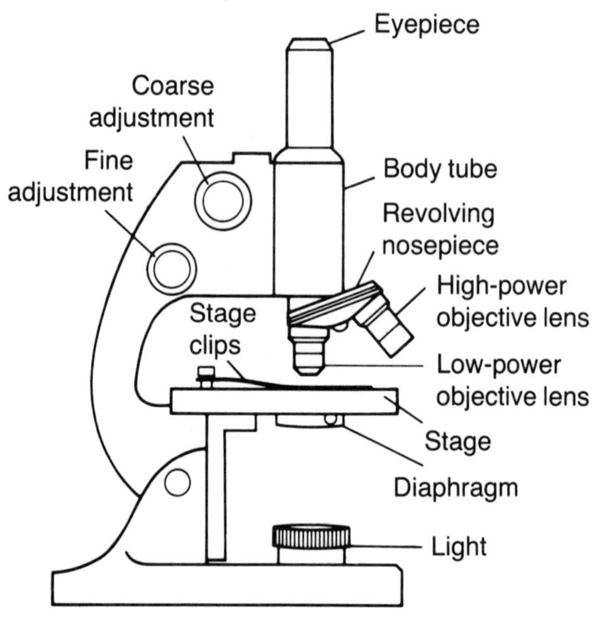

Care of the microscope

1. Carry the microscope upright. Use two hands. Hold the microscope by the arm with one hand, and use the palm of the other hand to support the base of the microscope.

2. When using the coarse-focus knob, always move the lens in a direction away from the stage.

3. Use only lens paper to clean the microscope lenses and the mirror, if your microscope has a mirror.

4. If your microscope has a substage illuminator (electric light), leave it on only when you are viewing a specimen under the lens.

5. If you replace a slide, repeat the procedure for focusing from the beginning.

6. If you spill any water or stain on the stage, immediately use a paper towel to clean up the spill.

7. Always store the microscope with the barrel lowered. Always cover the microscope with its dust cover when you store the microscope.

Using the compound microscope

1. Place the slide on the stage of the microscope. Be sure that the object you want to observe is over the opening in the stage.

2. Secure the slide with the stage clips.

3. Set the diaphragm to its widest opening. While looking through the eyepiece, adjust the light so that you see a bright field of view.

4. Check to see that you have the nosepiece set on the low-power objective. If not, turn the nosepiece until the low-power objective clicks into place.

5. To begin focusing, watch from the side as you turn the coarse-adjustment knob until the objective lens is very close, but not touching, the slide.

6. Then, looking through the ocular lens of the eyepiece, slowly raise the objective lens by using the coarse-adjustment knob. Continue raising the objective lens until the specimen comes into view.

7. Once the specimen is in view, use the fine-adjustment knob to bring the specimen into sharp focus.

8. Adjust the opening of the diaphragm to give contrast and detail to the specimen.

9. If you want to view the specimen under the high-power objective, turn the nosepiece until the high-power objective clicks into place. Use the fine-adjustment knob to bring the specimen into focus. Caution: *Do not use the coarse-adjustment knob when viewing under high power. You may damage the slide or the lens.*

10. You may have to readjust the opening of the diaphragm under high power. Use different settings of the diaphragm to find the clearest image in your specimen.

Making a wet-mount slide

Wet-mount slides are used to examine plant and animal cells under the microscope. The specimen being viewed is prepared in water on the slide. Onion membrane, cheek, and *Elodea* cells are often prepared as wet-mount slides.

To prepare a wet-mount slide, obtain a clean glass microscope slide and cover slip. Gently place the specimen in the center of the slide. *Note: Be sure that your specimen is thin enough for light to pass through it.* Use a medicine dropper to place one drop of water on the specimen.

Place one edge of the cover slip against the side of the drop of water. Use a dissecting needle or your fingers to support the other side of the cover slip. Gently, lower the cover slip over the water drop and the specimen until the cover slip lies flat. Check to be sure that there are no air bubbles trapped beneath the cover slip. If air bubbles are present, gently tap the cover slip with a pencil eraser or gently press down on the cover slip with the dissecting needle.

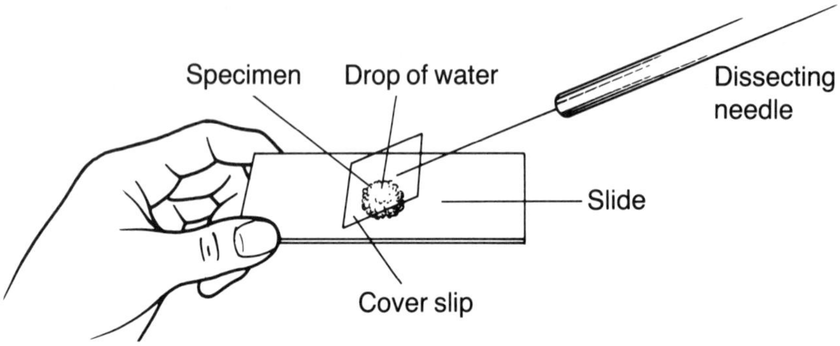

Staining cells

To stain cells, follow the procedure for making a wet-mount slide. Then, place a small drop of stain at one edge of the cover slip. The stain can be drawn under the cover slip by touching a piece of paper towel or lens paper to the opposite side of the cover slip. The cells will then be stained. Some stains that you may use are Lugol's iodine solution and methylene blue.

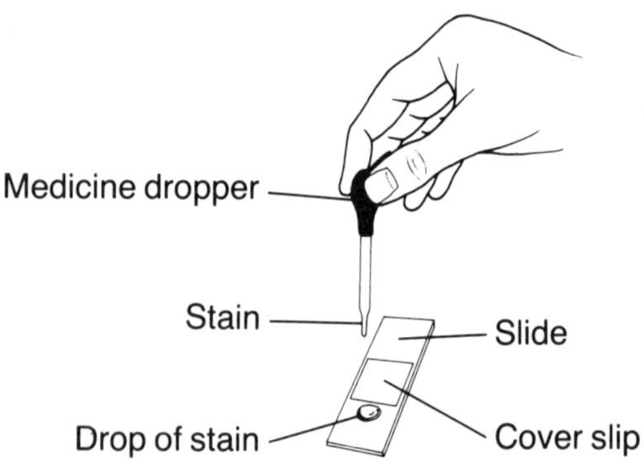

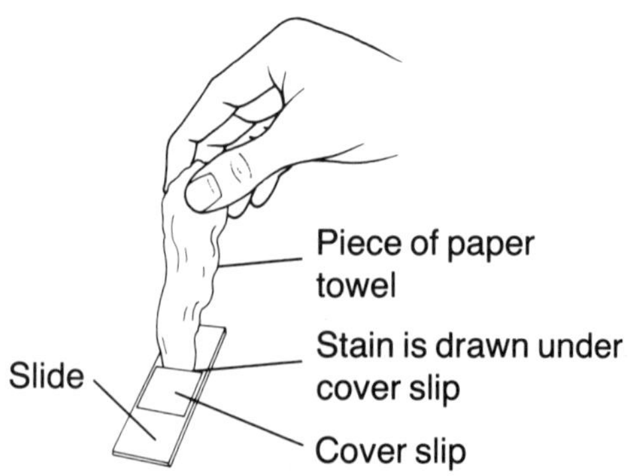

RESEARCH LAB

Growing Plants

1

Purpose: To study the effects of the best growing conditions on a particular plant

Materials you may use

seeds (choose one kind):
- onion seeds
- radish seeds
- corn seeds
- bean seeds

plastic growing cups

water

kind of soil (try different mixtures):
- garden soil
- peat moss
- sand
- vermiculite
- potter's clay
- silt

soil treatments (choose one):
- lime
- fertilizer

Background information: There are scientists who do research on growing plants. They try to determine the best soil and growing conditions for each kind of plant, so that the farmer will have a good harvest and will plant the best crops for the land. In this research project, you will choose a kind of crop and determine the best growing conditions for that particular plant.

In all experiments your *responding variable* will be growth rate. Experimental variables could be soil composition, exposure to sunlight, water temperature, air quality, etc. All except one should remain *constant* in both tests. One variable will be *manipulated,* or changed, in each test.

Techniques you may use:

1. ***Choosing seeds.*** Focus on one kind of plant. You can try others later.
2. ***Choosing soil.*** Select three very different soils for your seeds—at least one should drain freely, and at least one should hold water.
3. ***How to set up soil experiment.*** Choose one of the soil treatments. Use it as the product label states. You will need two cups for each soil type chosen—one cup to receive the soil treatment and the other to serve as a control.
4. ***How to use sunlight as a controlled variable.*** Fill your cups most of the way with soil, and plant several seeds in each. Water the cups when the soil feels dry. Leave all cups in a sunny or brightly lighted place. All cups should receive the same lighting.
5. ***How to write a soil report.*** After the plants have begun to grow, observe them carefully, and make a report. Your report should include all the conclusions that you can draw from your research.

Research strategy: In order to solve the problem described in the purpose, you need to follow each of the steps outlined below. Use a separate sheet to write your procedures and collect your data.

1. Establish a hypothesis.

In your hypothesis, include an assumption about the problem and a prediction that can be tested using the materials and techniques listed on the preceding page. Write your hypothesis on a separate sheet of paper.

2. Decide how to test the hypothesis.

Plan how you will test your hypothesis. Remember that you may use only the materials found in the list on the preceding page. You may use any of the techniques described, as well as common laboratory techniques, such as using a microscope. You may not need all of the materials and techniques given. Be sure to test only one variable in your experiment and to include a control. Record the steps you plan to follow on a separate sheet.

3. Perform the experiment.

Set up the experiment and gather data. Draw a chart, table, or diagram to record your data on a separate sheet.

4. Analyze the data.

Figure out what the data mean. If helpful, draw a graph on a separate sheet.

5. Draw a conclusion.

Determine whether the results of your experiment support your hypothesis or not. Answer the following questions in drawing your conclusion.

Write your answers on a separate sheet of paper.

1. Write your hypothesis.
2. Did your data support your hypothesis or not? Why or why not?
3. What can you conclude from this experiment?
4. What additional tests would you perform?

RESEARCH LAB

The Effect of Salt on the Growth of Bacteria

2

Purpose: To study the effects of salt water on bacteria and to discover the significance of this information

Materials you may use

pond water
at least 4 test tubes and a holder
masking tape
graduated cylinder
diluted methylene blue dye
medicine dropper
table salt

Background information: Bacteria are among the simplest of creatures. They live by decomposing organic matter. Your goal is to determine how salt affects the behavior of bacteria.

In all experiments your *responding variable* will be the difference in bacterial behavior. Variables remaining *constant* in all tests could be temperature, water volume, food substances, etc. The variable being *manipulated* is the amount of salt added.

Techniques you may use:

1. ***How to make "sea water."*** Add 35 g of table salt to one liter of aged tap water (aged tap water has been left standing for 24 hours).

2. ***Setting up a controlled variable, or control.*** During an experiment it is often necessary to set up a control. The control is needed to compare with the experimental results. The experimental setup differs by only one factor from the control.

3. ***Bacteria and methylene blue dye.*** Bacteria act on methylene blue dye, causing it to change from a blue to a colorless liquid (20 drops for every 50 mL of sample is recommended).

Research strategy: In order to solve the problem described in the purpose, you need to follow each of the steps outlined. Use a separate sheet to write your procedures and collect your data.

1. Establish a hypothesis.

In your hypothesis, include an assumption about the problem and a prediction that can be tested using the materials and techniques listed. Write your hypothesis on a separate sheet.

2. Decide how to test the hypothesis.

Plan how you will test your hypothesis. Remember that you may use only the materials found in the list on the preceding page. You may use any of the techniques described, as well as common laboratory techniques, such as using a microscope. You may not need all of the materials and techniques given. Be sure to test only one variable in your experiment and to include a control. Record the steps you plan to follow on a separate sheet.

3. Perform the experiment.

Set up the experiment and gather data. Draw a chart, table, or diagram to record your data on a separate sheet.

4. Analyze the data.

Figure out what the data mean. If helpful, draw a graph on a separate sheet.

5. Draw a conclusion.

Determine whether the results of your experiment support your hypothesis or not. Answer the following questions in drawing your conclusion.

Write your answers on a separate sheet.

1. Write your hypothesis.
2. Did your data support your hypothesis or not? Why or why not?
3. What can you conclude from this experiment?
4. What additional tests would you perform?

RESEARCH LAB

The Effect of Pollution on Seeds

3

Purpose: To study the effects of pollution caused by detergents on the germination of pinto bean seeds

Materials you may use

2 or more beakers (100 mL)
liquid detergent
pinto-bean seeds
plastic bags
water
tape
spoon or rod for stirring

Background information: Plants are affected by the pollution in their environment. Germination is a very vulnerable stage in the plant life cycle. Water pollution can slow down or stop germination. Your goal in this laboratory is to determine the effects of pollution on seed germination.

In all experiments, your *responding variable* will be the seed germination rate. Variables remaining *constant* in all tests could be quantity of water, sunlight, temperature, etc. The variable being *manipulated* could be the amount of detergent added.

Techniques you may use:

1. ***How to soak beans for germination.*** 24 hours is the recommended soaking time for pinto beans in preparation for germination.

2. ***How to set up seeds for germination.*** Place soaked seeds between two moistened paper towels. Put beans and towels in a plastic bag, and seal the bag.

3. ***How to prepare soaked beans for germination.*** Beans should be rinsed in fresh water before putting them on moist towels.

4. ***Setting up a controlled variable, or control.*** During an experiment, it is often necessary to set up a control. The control is needed to compare with the experimental results. The experimental setup differs by only one factor from the control.

5. ***Time needed for germination.*** Seeds should germinate in 48 hours.

6. ***Comparing results.*** Germination can be reported in terms of the percentage of seeds that germinated.

7. ***Making polluted water.*** Polluted water can be made by mixing various amounts of water with liquid detergent. Different concentrations of detergent can be tried.

Research strategy: In order to solve the problem described in the purpose, you need to follow each of the steps outlined below. Use a separate sheet to write your procedures and collect your data.

1. Establish a hypothesis.

In your hypothesis, include an assumption about the problem and a prediction that can be tested using the materials and techniques listed on the preceding page. Write your hypothesis on a separate sheet.

2. Decide how to test the hypothesis.

Plan how you will test your hypothesis. Remember that you may use only the materials found in the list on the preceding page. You may use any of the techniques described, as well as common laboratory techniques, such as using a microscope. You may not need all of the materials and techniques given. Be sure to test only one variable in your experiment and to include a control. Record the steps you plan to follow on a separate sheet.

3. Perform the experiment.

Set up the experiment and gather data. Draw a chart, table, or diagram to record your data on a separate sheet.

4. Analyze the data.

Figure out what the data mean. If helpful, draw a graph on a separate sheet.

5. Draw a conclusion.

Determine whether the results of your experiment support your hypothesis or not. Answer the following questions in drawing your conclusion.

Write your answers on a separate sheet.

1. Write your hypothesis.
2. Did your data support your hypothesis or not? Why or why not?
3. What can you conclude from this experiment?
4. What additional tests would you perform?

A C T I V I T Y

Ecology Field Trip

1

Purpose: To provide an opportunity to review firsthand some of the important concepts of ecology and to sharpen library research skills

Time required: 2 hours outside of class

Materials

notebook
day pack
textbook
pen or pencil
binoculars if available

Procedure: Write your answers on a separate sheet.

1. After studying the material on ecology presented in Chapters 3 and 4, take a short hike in a neighboring field, forest, or park (preferably not a city park). In the field review the following concepts:

a. Biomes
b. Food chains
c. Food webs
d. Niche
e. Habitat
f. Ecological pyramids
g. Energy flow
h. Nutrient cycling
i. Limiting factors

See text glossary for definitions.

2. Begin by defining the term. Then, find examples. Take notes and draw diagrams of your observations. Jot down any impressions or new insights you have.

3. At home, visit a library to research your biome. Some facts to look for are the major plants and animals, average temperature, the amount of precipitation, precipitation patterns (for example, the percentage of precipitation falling as rain and snow and also the seasonal precipitation).

4. Also look at ways your biome has been altered. For example, what foreign species have been introduced? How have they affected it? (You may have to call your Wildlife Department, contact a wildlife biologist, or talk to hunters and anglers to get information on introduced species and their benefits and costs.) What other ways has your biome been altered? What major resources are produced in your biome?

5. When you are finished, write a 500-word report on your biome, including the pertinent discoveries in the field and the library.

Report format: Answer the questions above, then write report separately.

ACTIVITY 2

Family Size and Population Growth

Purpose: To observe the effect of family size on population growth in the U.S. by comparing growth rates based on 2- and 3-child families

Time required: 1 period

Materials

graph and information provided

Figure 1

United States population projections based on 2- and 3-child families

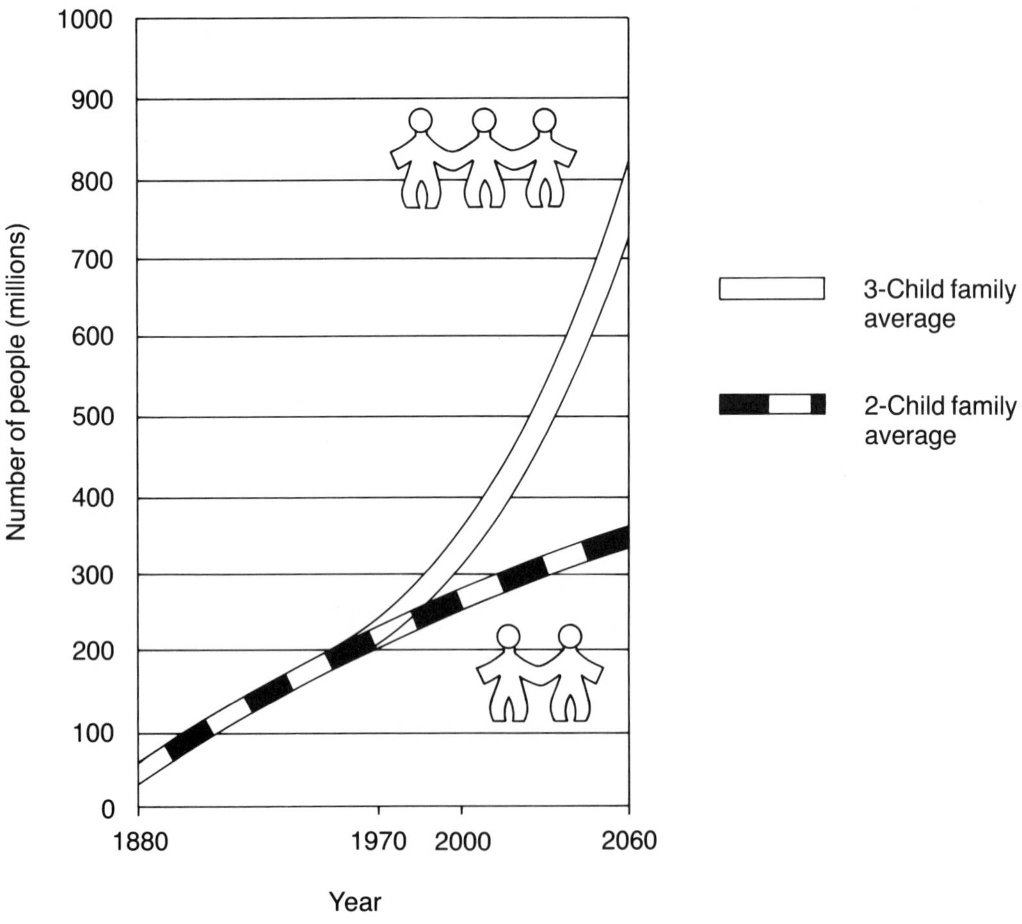

Name ________________ Class ________ Date ________

Background Information: Most of you will predictably live through a large part of the 21st century. During this time you will have children, and your children will have children. In fact many of you will live until your grandchildren begin to have children. This fourth generation and those who survive from the 3 previous generations will form the population of the United States at that time. Our decisions to marry, have children, or move obviously have important consequences for our own lives as well. The sum of these decisions also has a major impact on the whole society today and many years from now.

An example is *family size.* If one family with 2 children decides to have a third child, the choice is certainly important for each member of that family. But if we think of all families making the same decision, the results in numbers alone are surprising as you can see by studying the graph in Figure 1. If all families in the United States have an average of 2 children (and immigration remains at the current level), in 100 years our population will grow from the current level of 230 million to over 350 million. If, however, families have an average of 3 children, the population will increase to nearly 1 billion people in 100 years! That is a difference of 650 million people in only 100 years. We can see how the growth of population in our country depends on choices individual parents will make in the next few years.

Interpreting a graph: Does the graph in Figure 1 properly represent the information stated in the paragraph above? The y axis (the vertical line at the left) represents the number of people in millions. The x axis (horizontal line at bottom) represents specific years. The lines are drawn through several plots. For example in 1970 both plots are at the 200-million line. In the year 2000 the 3-child family is slightly above the 300-million line. Where is the plot for the 2-child family?

Questions: Write your answers on a separate sheet.

1. If families from 1970 on had an average of 2 children, what would the population be in the year 2060? What would the population be if families had an average of 3 children?

____

____

2. What would be the effects of the 3-child family (compared to the 2-child family) on the demand for water, energy, education, and space in the national parks?

____

____

____

____

Name ______________________ Class ____________ Date ____________

3. What are some other decisions that you will make in your lifetime that may affect the lives of your own and other people's children and grandchildren?

Type of car driven, whether or not you go to college, whether or not you get married right out of high school, the type of house you build or purchase, the type of recreation you choose.

4. On a separate sheet, make a graph showing the family sizes of your class members. Find out how many come from 1-child families, 2-child families, etc. Represent the number of children in each family along the x axis (1, 2, 3, etc.) and the number of students coming from each-size family along the y axis.

Report format: Answer questions as shown and plot the graph as described.

ACTIVITY

The Roots of Hunger

3

Purpose: To research the 8 contributing factors to world hunger and how each can possibly be overcome

Time required: 2 class periods: 1 for group study, 1 for class discussion

Materials

textbook

recent newspaper, magazine and/or journal articles on hunger in Africa

Procedure: Write your answers on a separate sheet.

1. Students gather into 8 working groups. Each one will research 1 element of the model shown here: population, per capita consumption, politics, psychology, etc. Chapters 7, 8, and 9 and suggested readings at the end of each, and articles from magazines and journals listed in Appendix B may prove helpful. Study hunger in Africa and determine what effects the factors presented in the accompanying figure have on hunger and starvation. For example, in what ways does population affect hunger? And how have politics (or poor public policy) made matters worse? Have non-political relief organizations/functions helped? Some connections may not be as clear as others. For instance, how has technology affected hunger?

2. Present the findings of your research in a class discussion. List the findings on the board and then decide by vote which factors seem to be the most important ones. Draw up a plan to reduce hunger in Africa.

Figure 1

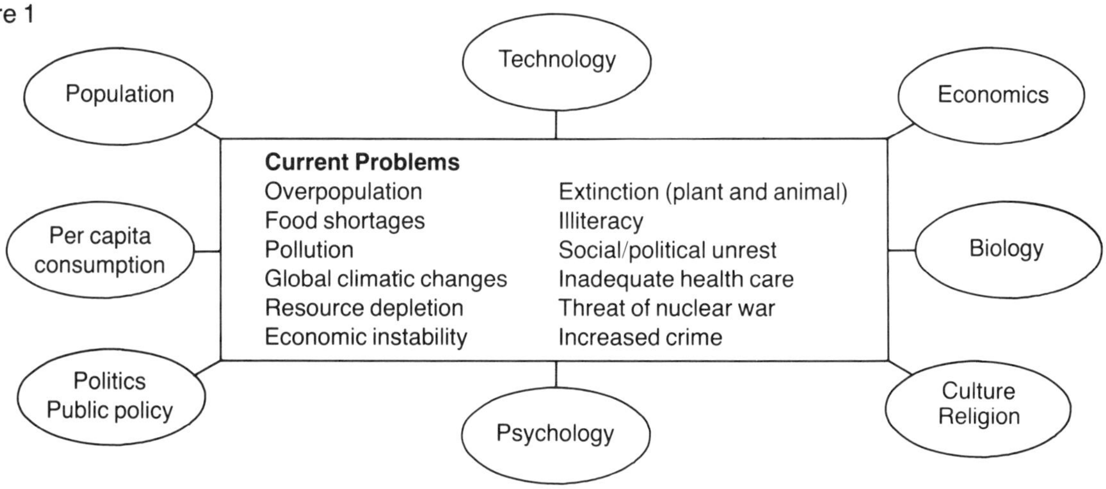

3. Each group should consider one of these points:

a. ***Population:*** What effect do overpopulation and rapid population growth have on the task of supplying food?

b. ***Per capita consumption:*** This one won't yield much information. But try to determine if excessive per capita consumption in developed nations affects developing nations. Can you think of any other ways per capita consumption in Africa and the developed nations affects hunger?

c. ***Politics and public policy:*** This is one that will yield a great deal of information. How have agricultural policy and population policy affected hunger? Have internal political battles affected the food supply in Ethiopia? How have developed nations affected food production in the developing nations?

d. ***Psychology:*** Have Third World nations become dependent on the developed nations for food aid? What psychological factors contribute to overpopulation, and hence food shortage?

e. ***Culture:*** Have cultural barriers impeded the development of a more efficient agricultural system? Have cultural differences contributed to political differences that cripple governments and make other policies weaker?

f. ***Biology/Natural Causes:*** What effects have droughts had? What effects has the Green Revolution had? What is the African soil like?

g. ***Economics:*** How has poverty impaired the task of feeding the world's hungry people? How has the price of oil affected food?

h. ***Technology:*** Has modern technology helped African nations to reduce hunger?

Report format: Answer questions as shown.

ACTIVITY

Climatograms

4

Purpose: To relate climate to the plant and animal communities of different areas by plotting climatograms of biomes

Time required: 1 or more periods

Materials

graph paper with ¼″ squares

Background information: A biome is a major ecological community having characteristic plants and animals. Biomes are closely associated with 2 climatic factors, temperature and precipitation. In most biomes, temperature and precipitation vary throughout the year. These variations are plotted on graphs called *climatograms.* Climatograms provide a rough idea of the climatic conditions in a particular area, and help scientists understand the demands placed on organisms in the diverse communities on earth.

Figure 1 is a climatogram from a middle latitude deciduous forest. The bar graph shows average monthly precipitation, while the line graph illustrates average monthly temperature.

Figure 1

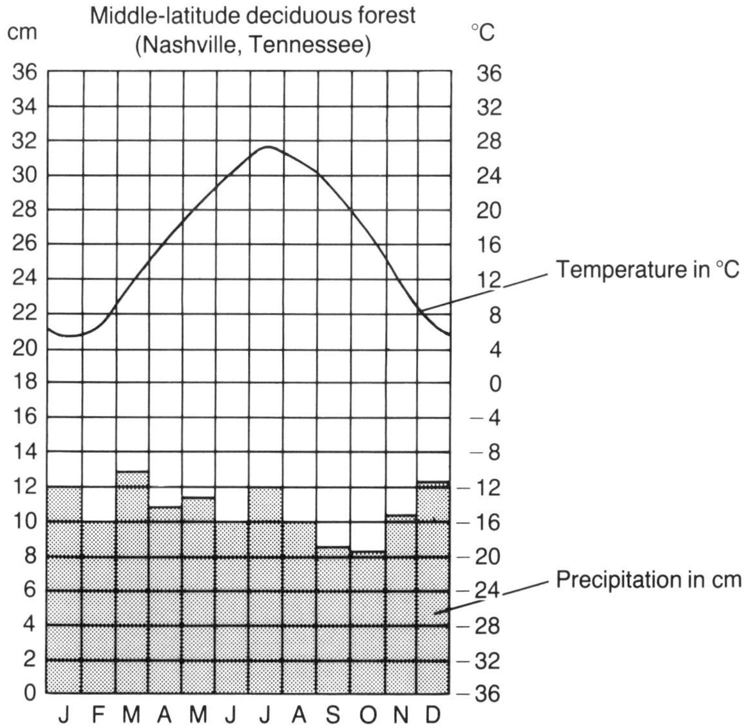

Name ______________________ Class ______________ Date ______________

Procedure:

Student climatograms should follow the format of Figure 1.

1. From the data below, plot a climatogram like Figure 1 for each of the following biomes. Be sure to label each climatogram.

1. Tropical Rain Forest (Manokwari, New Guinea)

(Months)	J	F	M	A	M	J	J	A	S	O	N	D
T (temp. °C)	27	26	27	27	27	27	27	27	26	27.2	27.2	25
P (prec. cm)	28	27	33	28	19	21	15	13	13	11	17	27

2. Taiga (Anchorage, Alaska)

(Months)	J	F	M	A	M	J	J	A	S	O	N	D
T	−11	−8	−4	0	6	12	14	12	8	2	−6	−10
P	2	1	1	1	1	3	4	7	7	5	3	2

3. Arctic Tundra (Barrow, Alaska)

(Months)	J	F	M	A	M	J	J	A	S	O	N	D
T	−26	−28	−26	−20	−8	0	4	3	−2	−10	−18	−24
P	.5	.5	.5	.5	.5	1	2	2	1.5	1	.5	.5

4. Grassland (Lawrence, Kansas)

(Months)	J	F	M	A	M	J	J	A	S	O	N	D
T	−2	2	6	12	18	24	26	26	22	14	6	0
P	3	3	5	9	13	12	9	11	8	6	5	4

5. Northern Desert (Reno, Nevada)

(Months)	J	F	M	A	M	J	J	A	S	O	N	D
T	−1	2	6	8	12	16	20	14	14	10	4	0
P	3	3	1.5	1	1	1	.5	.5	.5	1	1	2

6. Chaparral (Santa Monica, California)

(Months)	J	F	M	A	M	J	J	A	S	O	N	D
T	11.7	11.7	12.8	14.4	15.6	17.2	18.9	18.3	18.3	16.7	14.4	12.8
P	8.9	7.6	7.4	1.3	1.3	0	0	0	0.3	1.5	3.6	5.8

2. After completing the climatograms, answer the following questions.

a. Which biome has the lowest temperatures throughout the year? Which has the most regular rainfall? Which has the most irregular rainfall pattern?

Arctic tundra, tropical rain forest, chaparral and grassland.

b. What similarities are there between the arctic tundra biome in Barrow, Alaska, and the desert biome in Reno, Nevada?

Low precipitation.

Name ______________________ Class ____________ Date ____________

c. What is the major climatic difference between the grasslands biome and the northern desert biome?

Rainfall.

d. On text pages 65 through 70, read the description of the plants and animals in several biomes. Choose one biome and list several features which show that organisms in it are adapted to the climate of their area.

(Various answers.)

Report format: Hand in climatograms and answers to questions.

ACTIVITY

Water Research

5

Purpose: To promote a better understanding of local water supplies and ways to meet increasing water demands

Time required: 1 afternoon and evening after school

Materials

notebook	pen or pencil

Procedure: Write your answers on a separate sheet.

1. Contact your local water department. Ask their public information officer to provide you with information regarding your community's current water system. Find out:

a. The local population served by the system;

b. How much water is used by each sector (farmers, homes, commerce);

c. How water is used in households (that is, the percentage used for washing, drinking, lawn watering, etc.);

d. Projected increases in demand in each sector;

e. Where the water comes from.

2. Once you have a picture of the current water supplies and water demands, devise a plan to meet future needs. Look at the following options individually and collectively:

a. Population control;

b. Conservation;

c. Water recycling;

d. New water supplies (wells or surface water).

For each of these options make a list of their pros and cons. For example, conservation may be much cheaper than building a new dam, but would require new laws or voluntary cooperation. Try to determine how much water each option could provide and the cost. Water department engineers and economists may be willing to help you in this regard.

3. When you have finished your study, draw up a brief document outlining the strategy that best meets your community's needs. What are the major environmental impacts of each option?

Report format: Answer questions as shown.

ACTIVITY 6

Water Quality

Purpose: To learn to perform water-quality tests and to interpret their results

Time required: 3 or more periods; field trip to collect water samples

Materials

water-quality test kit

water samples from natural and treated sources

Name ______________________ Class ____________ Date ____________

Procedure:

Test each sample as described below, and record your results as shown in Figure 1. It may be necessary to repeat some tests to determine how consistent your results are.

Figure 1 Data Table

Test	Results	Recommended Standards
Water Hardness		Less than 300 ppm
pH		6.5–7.8
D.O.		6–15 ppm
Nitrate		Less than 45 ppm
Phosphate		0.1 ppm or less
TDS		Less than 500 ppm
Total Coliforms		
Fecal Coliforms		

Source of sample: ____________

Date collected: ____________

Date tested: ____________

1. Water hardness

Water hardness may be caused by the natural accumulation of salts of calcium and magnesium. Water absorbs these elements when it comes in contact with soils and rock, or industrial pollution. Excessive hardness is undesirable, as it may cause dry skin, yellowing of clothing, and the clogging of plumbing systems. Some minerals are harmful to health.

Follow the instructions found with the test kit to determine the hardness of your water sample.

Hardness of water sample______.

2. Water pH

The pH value determines the acidity or alkalinity in water. A pH value of 7.0 is neutral (neither acidic or basic). A pH value of below 7 indicates acidic water and a pH reading of above 7 indicates alkaline water. Most species must live in an aquatic environment at or near 7.0. Any fluctuation in pH may cause death of the aquatic organisms. Follow the test kit instructions to determine the pH of your water sample.

pH of water sample______.

3. Dissolved Oxygen (D.O.)

This is probably the most important test in determining the water quality in a lake or stream. It determines the suitability of the water for fish and other aquatic organisms.

Follow the instructions found in the test kit to determine the dissolved oxygen concentration of your water sample.

Dissolved oxygen concentration______.

4. Nitrate

Nitrate may result from runoff of livestock feedlots or heavily fertilized fields. Nitrates in excess concentrations may cause large growths of algae which ultimately lead to deterioration of water quality.

Follow the directions found in the test kit to determine the nitrate concentration of your water sample.

Nitrate concentration______.

5. Phosphate

Like nitrates, high concentrations of phosphates in water may cause excessive algae growth which leads to a decrease in oxygen available to aquatic animals, especially fish. And in the fall, when the algae die, it is consumed by oxygen-consuming bacteria in aquatic systems.

Follow the directions found in the test kit to determine the phosphate concentration of your water sample.

Phosphate concentration______.

6. Total Dissolved Solids (TDS)

TDS is a combination of all salt and mineral ions dissolved in water. High TDS can affect the osmotic pressure of water and thus be detrimental to aquatic life. Follow the directions found in the test kit to determine the TDS concentration of your water sample.

TDS concentration______.

7. Coliform bacteria

Coliform bacteria are a harmless group of bacteria found in soil and the intestines of both animals and humans. The presence of coliform bacteria in a water sample indicates pollution by sewage.

Follow the directions found in the test kit and determine if coliform bacteria are present in your water sample.

Total coliform/100 mL of water______.

Fecal coliform/100 mL of water______.

Name ______________________ Class ______________ Date ______________

After all the tests have been done, each group should enter its data in a table as shown in Figure 2. Circle any value that exceeds the recommended standards, then answer the following questions on a separate sheet.

Figure 2 Water Quality Class Results

	Hardness	pH	DO	Nitrate	Phosphate	TDS
Recommended Values	< 300 ppm	6.5–7.8	> 6 ppm	< 45 ppm	≤ 0.1 ppm	< 500 ppm
Source of Sample						
1.						
2.						
3.						
4.						
5.						
6.						
7.						
8.						
9.						
10.						
11.						
12.						

1. List the sample sites that exceed water quality standards, and the category in which they exceed these standards.

2. Which water quality standard was exceeded most often? What could be the reason for this?

3. For one sample site, note what tests show excesses above recommended standards and suggest what the environmental and human consequences of these excesses might be. Suggest some possible causes and possible ways to correct these problems. Be specific.

Report format: Hand in data tables and answers to questions.

ACTIVITY

Electrical Energy in the Home

7

Purpose: To determine how much electrical energy is used at home, and to devise and test a plan for reducing our electrical energy use

Time required: 2 or more periods

Materials

old electrical bills brought in by students

calculators will be helpful

Procedure: Write your answers on a separate sheet.

1. Everyone in the class should determine the number of kilowatt hours (1,000 watts) of electricity used (for a 1-month period) in his/her home. Do this by bringing electric bills for at least 6 months to school in order to calculate a monthly average of kilowatts used. (This information appears on the bill under "total kilowatts" used.)

2. Next, make a list of the specific uses of electricity in your home, starting with the highest user and ending with the lowest. Table A, which follows Figure 4, shows the major uses on a year-round basis for the U.S. This shows the wattage, annual kilowatt hour consumption and annual cost for many different electrical appliances. The annual kilowatt hour consumption is calculated in this way:

Figure 1

$$\frac{\text{kwh}}{\text{yr}} = \frac{\text{watts}}{\text{hours of use}} \times \frac{\text{hours used}}{\text{yr}} \times \frac{1\ \text{kw}}{1000\ \text{watts}}$$

Divide the answer by 12 to determine the monthly consumption. The annual cost is determined by multiplying the KWHR consumption by the cost per kilowatt hour, shown on the electric bill. In the table we use $0.07 per KWH.

If your appliances have different average wattages from those given, calculate the consumption as shown above. You will have to estimate hours of use; for example, the refrigerator is running only some fraction of the day. Only the electric clock runs all day and night.

Calculate the electrical consumption of your appliances, and determine the total average consumption for your home. See how this compares to the actual kilowatt hours reported on the electric bill.

3. Using Figure 2 as a guide, devise a plan for your household to consume 15% less electricity in the next month. Convince your family to help you carry out the plan for reducing electricity consumption. After

1 month, check the electric bill to see whether you succeeded. Evaluate the benefits as well as any costs of the project. Will you continue to conserve electricity?

Figure 2

Proportional use of electrical energy in the home yearly

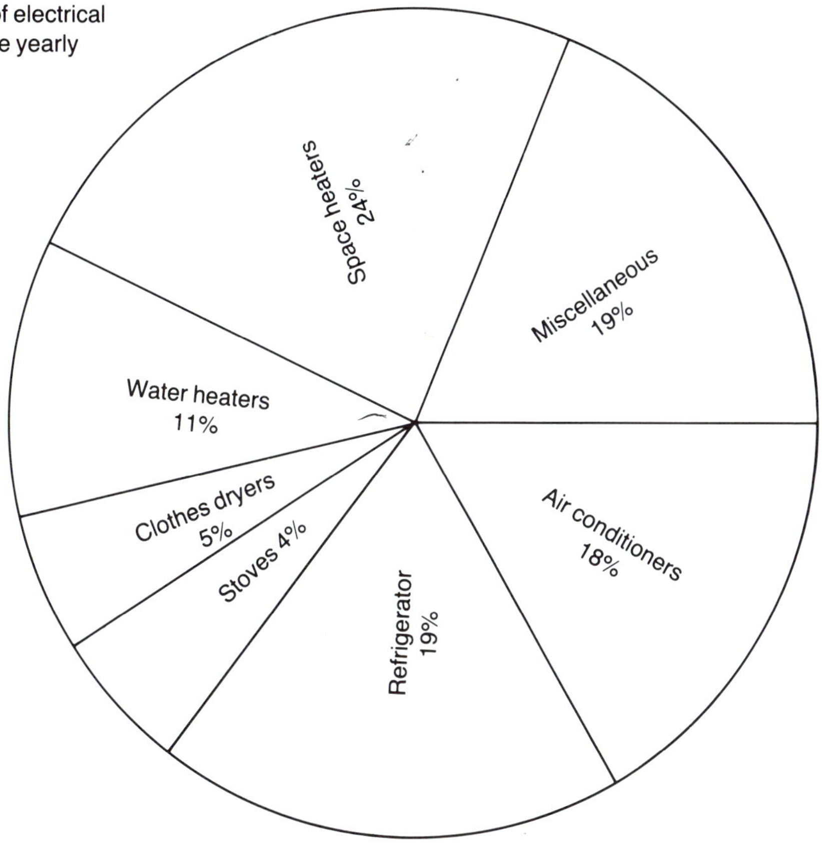

Report format: Turn in a data sheet showing your calculations for Procedure 2. Write a short report on your results in Procedure 3.

Other calculations: To determine the hours of use estimated for each appliance listed in Table A (for example, for the range):

Figure 3

$$\frac{\text{hours used}}{\text{yr}} = \frac{1{,}175 \text{ kw/yr}}{12.2 \text{ kw/hours used}} = 96 \text{ hours per year}$$

To determine KWHR consumption for appliances of different wattage from those given (assuming the same hours of use):

Figure 4

e.g. if your range uses only 10,000 watts

$$\frac{10{,}000 \text{ watts}}{12{,}200 \text{ watts}} \times (1{,}175 \text{ kwhr cons.}) = 963 \text{ kwhr cons.}$$

if your range uses 13,000 watts

$$\frac{13{,}000 \text{ watts}}{12{,}200 \text{ watts}} \times (1{,}175 \text{ kwhr cons.}) = 1{,}250 \text{ watts}$$

Extend the project by having students consider what would happen if everyone reduced their energy consumption.

Table A 1982 Wattage and Estimated Kilowatt-Hour Consumption for Electrical Appliances

Appliance	Average Wattage	Annual KWHR Consumption	*Annual $ Cost
Air conditioner (room)	1,566	1,389	$ 97.23
Blanket, electric	177	147	10.29
Blender	386	15	1.05
Clock	2	17	1.19
Clothes dryer	4,856	993	69.51
Coffee maker	894	106	7.42
Dehumidifier	257	377	26.39
Dishwasher	1,201	363	25.41
Fan (attic)	370	291	20.37
Fan (circulating)	88	43	3.01
Fan (window)	200	170	11.90
Freezer (15 cu. ft.)	341	1,195	83.65
Freezer, frostless (15 cu. ft.)	440	1,761	123.37
Frying pan	1,196	186	13.02
Hair dryer	381	14	.98
Heater (portable)	1,322	176	12.32
Heater (infrared)	250	13	.91
Heating pad	65	10	.70
Hot plate	1,257	90	6.30
Humidifier	177	163	11.41
Iron (hand)	1,008	144	10.08
Lights—read wattage on light bulbs			
Mixer	127	13	.91
Oven, microwave	1,500	300	21.00
Oven, self-cleaning	4,800	1,146	80.22
Radio	71	86	6.02
Radio/record player	109	109	7.63
Range	12,200	1,175	82.25
Range, self-cleaning	12,200	1,205	84.35
Refrigerator/freezer (14 cu. ft.)	326	1,137	79.59
Refrigerator/freezer (frostless, 14 cu. ft.)	615	1,829	128.03
Sewing machine	75	11	.77
Shaver	14	1.8	.13
Sun lamp	279	16	1.12
Television (black & white) tube type	160	350	24.50
Television (black & white) solid state	55	120	8.40
Televison (color) tube type	200	660	46.20
Television (color) partial solid state	300	440	30.80
Toaster	1,146	39	2.73
Toothbrush	7	0.5	.04
Vacuum cleaner	630	46	3.22
Washing machine (automatic)	512	103	7.21
Waste disposal	445	30	2.10
Water heater (standard)	2,475	4,219	295.33
Water heater (quick-recovery)	4,474	4,811	336.77

* Annual cost based on electrical rate = 7¢/kwh

ACTIVITY

Recycling Waste Materials

8

Purpose: To identify the variety of items that we throw away that might be reusable

Time required: 1 period

Materials

1 large bulletin board or table

disposable items brought to class by each student

Background information: Numerous disposable, or throw-away, items are available to consumers, such as ballpoint pens, paper towels, napkins, facial tissues, cups and plastic eating utensils. Disposable items greatly increase the amount of garbage we throw away each year. Reusable materials could easily be substituted for disposables.

Procedure: Write your answers on a separate sheet.

1. Bring to class several examples of disposable items from home (be creative!).

2. Collect the items brought in and discuss what could be made out of each type of disposable item (paper, glass, plastic, rubber, etc.).

3. Construct on a bulletin board (or lay out on a flat surface) the collected items and identify them.

4. Determine the type of energy and materials that were used to make items. Discuss whether or not the items are biodegradable.

5. Consider the recycling possibilities of the collected items. Also discuss the reasons and needs for recycling waste materials.

6. Explain the relationship of recycling to energy supplies and consumption.

Report format: List 10–20 of the most common and disposable items used by students at school and at home. List recommendations for reducing the amount of materials thrown away and also how the items might be reused or recycled.

ACTIVITY

Solar Water Heater

9

Purpose: To demonstrate how water can be heated and circulated by solar power using a model system

Time required: 1 or 2 periods

Materials

glass cylinder (see Figure 1)
1 meter black rubber tubing
stand with clamp
short glass tube in 1-hole stopper to fit cylinder top
dye tablet or food coloring and pipette

Procedure:

1. Obtain a glass cylinder, rubber tubing and a support stand with clamp on it from your instructor.

2. Clamp the glass cylinder in an upright position in the support stand about 20 cm above the base.

3. Raise the free end of the tubing until the end of the glass tubing is a little above the top of the cylinder. Fill the cylinder *rapidly* with water. (See Figure 1.)

Figure 1

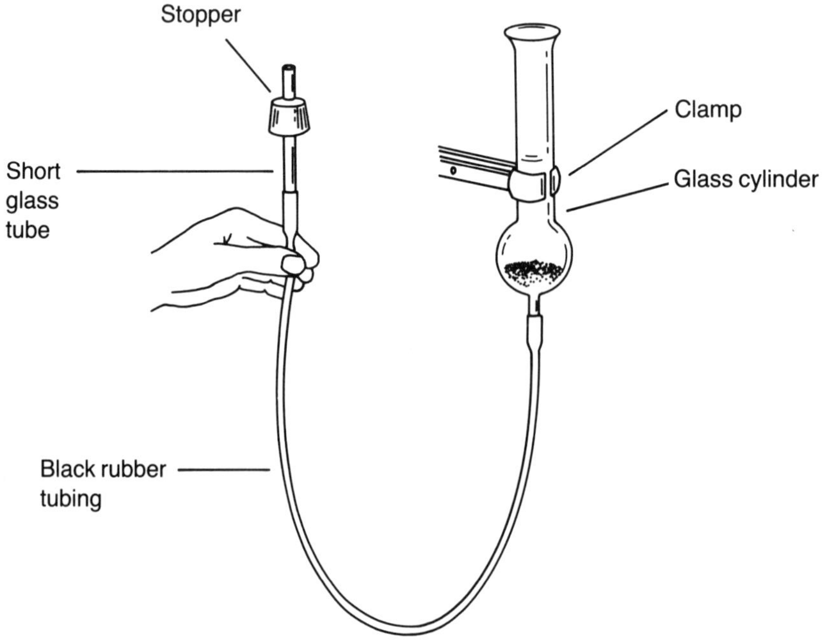

Name ______________________ Class ____________ Date ____________

4. Lower the stopper with tubing until the water level in the cylinder reaches the top of the bulb part of the cylinder. This will eliminate air from the system.

5. Raise the tubing until it is again level with the top of the cylinder, and refill the cylinder with water.

6. Now drop 1 red dye tablet into the cylinder; it will come to rest on a mesh platform in the cylinder bulb. Or, carefully introduce several drops of red food coloring into the bottom of the cylinder.

7. Place the stopper in the top of the cylinder, making certain the end of the short piece of glass tubing is below the surface of the water in the cylinder. When completed, your apparatus should look like Figure 2.

8. Make sure that there are no kinks in the rubber tubing.

9. Carefully place the apparatus in the sun for about 20 minutes, then answer the following questions.

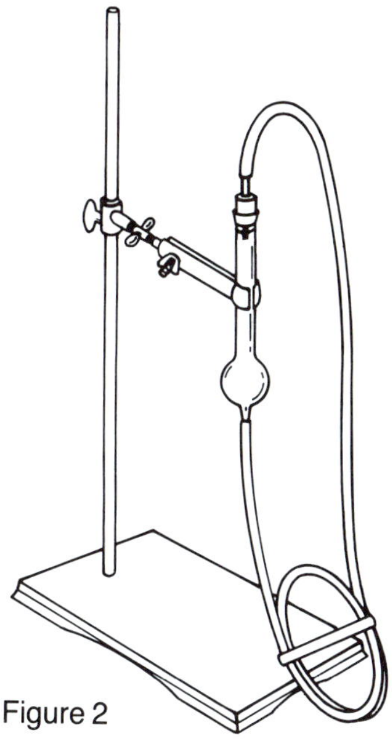

Figure 2

a. Trace the path of the red dye through the system. Why did the dye—and the water carrying it—move in this way?

The black tube heats up more than the rest of the system. Thus, water would rise into the cylinder, mixing the dye.

b. Why is black tubing used rather than light-colored tubing?

Being black, the tube absorbs more heat than if it were clear.

c. Would the system work better if it were placed under a glass cover? If so, why?

Yes. Glass would reduce heat loss.

Report format: Answer the questions above.

ACTIVITY 10

Solar Distillation

Purpose: To demonstrate the use of solar power to distill fresh water from salt water in a model system

Time required: 1 period

Materials

salt water
solar greenhouse (see Figure 1)
pipette or medicine dropper
black paper

Procedure:

The plastic cover and base shown in Activity 13 may be modified for use as a solar still. Any similar structure can be used provided it has some area for collecting the distillate uncontaminated by the salt water.

1. Obtain a container of salt water from the instructor.

2. Pour some of this salt water into the shallow pool in the center of the black base in the solar greenhouse. (Be careful that it doesn't overflow.)

3. Fold a piece of black paper lengthwise and stand it up in the pool, forming a tent-like cover. This acts as a wick and speeds up the solar distillation.

4. Place the plastic cover snugly over the base. Make sure that the spout in the cover is over the well in the base (see Figure 1).

Figure 1

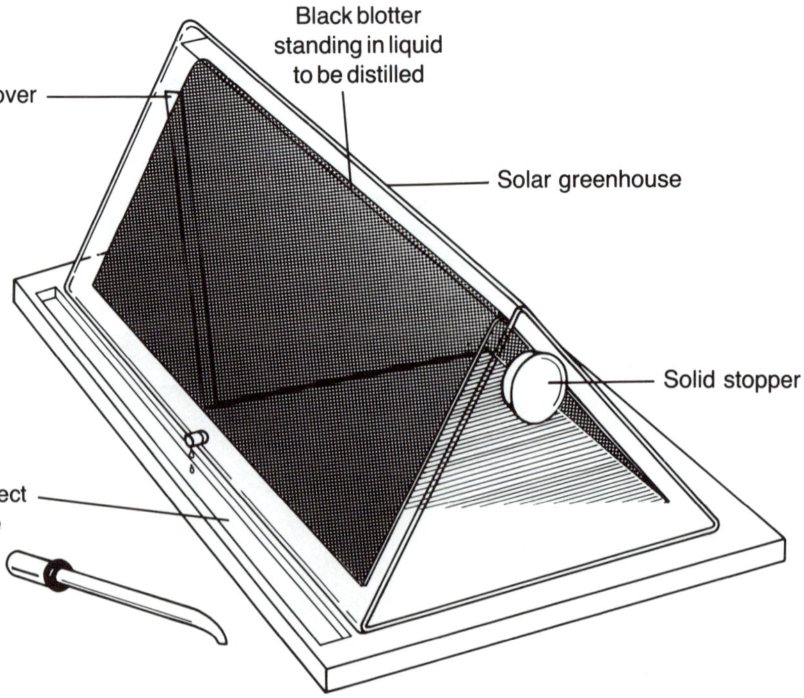

Name ____________________ Class ____________ Date ____________

5. Next insert the rubber stopper into the hole drilled in one end of the cover. Make sure that the plastic cover does not touch the black paper at any point.

6. Place the greenhouse in the sun for at least 30 minutes. During this time fresh water will accumulate in the well, and can be drawn off by a medicine dropper.

7. Compare the taste of the water from the distillate well to the taste of the original salt water, then answer the following questions:

a. Explain where the water in the distillate well came from.

The water in the distillate well came from the salt water pool. Sunlight energy evaporated the salt water, which then deposited on the inside of the enclosure and dripped into the well. When the water evaporates, it leaves behind the salts.

b. Compare distillation in your model system to the processes of the global water cycle.

The global water cycle is powered by sunlight. Water evaporates from oceans, freshwater supplies and so on. It becomes airborne and falls from the sky as rain or snow. Evaporation leaves behind impurities in both solar distillation and the global water cycle.

c. Suggest some changes that would be necessary to use a system like the model on a large scale.

Ready supply of salt water, pumps to transport salt and pure water, storage for salt and pure water, pipes to transport water to users, market for salt.

Report format: Answer the questions above.

A C T I V I T Y 11

Light Production

Purpose: To compare the amounts of light produced by fluorescent and incandescent lamps of the same wattage

Time required: 1 period

Materials

light meter
fluorescent and incandescent lamps—40-watt bulbs work well
meter stick

Background information: Recall that the wattage of light bulbs is a measure of the energy that they use.

Name ______________________ Class ____________ Date ____________

Procedure:

Relate the results of this activity to the second law of thermodynamics and to Activity 7.

1. Darken the room.

2. Using the light meter, measure the amount of light given off at a distance of 1, 2 and 3 meters by the 40-watt fluorescent lamp. Repeat for the incandescent lamp. Record your data on a separate sheet.

3. If there is a difference in the amount of light produced by the different types of lamps, suggest what is happening to the electrical energy which is not converted to light. What is the evidence for your suggestion?

Yes, the fluorescent light produces more light. The electrical energy not converted to light is converted to heat. The incandescent bulb feels much hotter.

Report format: Turn in data sheet and answer.

A C T I V I T Y

The Great Joule Robbery

12

Purpose: To demonstrate energy transformations and losses by building and operating a simple steam engine

Time required: 1 period

Materials

One-hole stopper	pinwheel
Florence flask or Erlenmeyer flask	support stand for blades
Bunsen burner and wire gauze	short piece of glass tubing
length of rubber tubing to fit glass tubing snugly	ring stand, ring, and 3 clamps

Background information: Engines transform various forms of energy into mechanical energy. At every transformation there is some loss of useful energy: this is the second law of thermodynamics. Since energy is measured in joules, we call this loss the joule robbery.

Figure 1

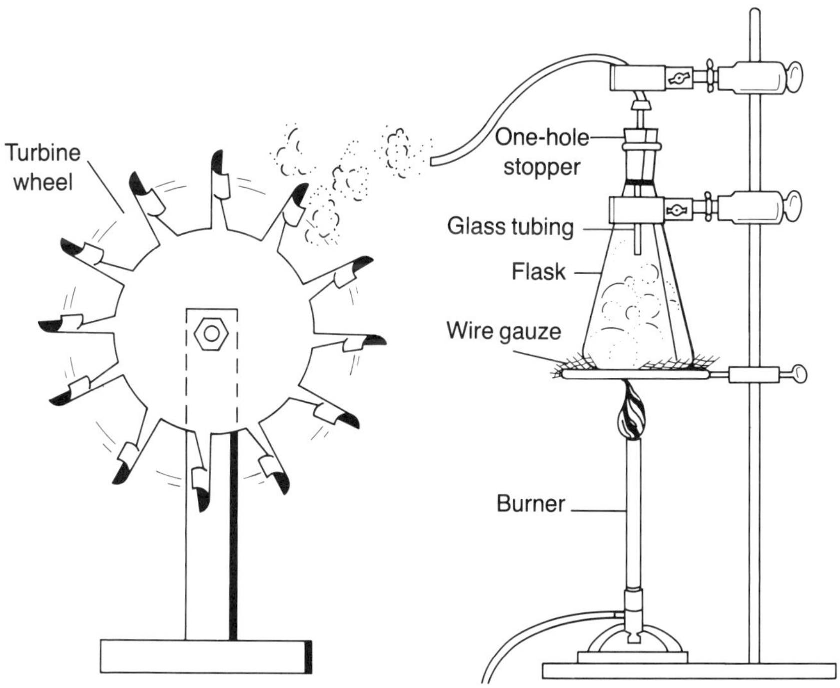

Procedure:

1. Set up the steam engine as illustrated and light the burner. Adjust the rubber tubing so the steam is directed toward the blades. Take care not to burn yourself with the steam.

Name ______________________ Class ____________ Date ____________

2. When you have enough steam to move the blades, observe the system and answer the following questions:

a. What form of energy powers the whole system? Where did this energy come from originally?

Natural gas. Underground deposits often associated with oil. Ultimately, the energy in natural gas came from sunlight.

b. What energy transformations occur in the system? List all the energy forms you can observe, including heat, light, mechanical energy and any others.

The energy in natural gas is converted into heat, light, kinetic energy of water molecules (steam), and mechanical energy.

c. What evidence do you observe of loss of *useful* energy at each energy transformation? (The useful energy is the energy which is available to power the next part of the system.)

Light given off by the burner is not useful energy. Heat given off by the flask and rubber tubing is also not useful energy in this system.

d. How could you make your model more efficient; that is, how could you change the model to reduce energy loss at each stage?

Design a more efficient burner and insulate the flask and rubber tubing.

e. You have constructed a turbine, an engine which includes blades turned by liquid or gas. Give some other examples of turbines.

Generator in an automobile. Electric generators in power plants and hydroelectric projects.

f. Observe common machines such as automobiles, power tools, stoves, electric and manual typewriters, etc. and note where energy is wasted: for example, any heat produced by a chain saw is wasted energy. Suggest ways that 1 or more of these machines might be made more energy-efficient.

(Various answers)

Report format: Answer questions as shown.

ACTIVITY 13

The Greenhouse Effect

Purpose: To demonstrate the greenhouse effect and suggest ways your observations can help you design solar collectors

Time required: 1 period

Materials

3 thermometers
1-hole stopper to hold 1 thermometer
black blotter or paper
white paper
plastic cover with base (see Figure 1)

Background information: Energy comes from the sun in the form of heat and light. When light strikes an object, some is reflected and some is absorbed by the object and converted to heat. Light can pass through glass (or plastic) and is converted to heat when it strikes light-absorbing materials. The heat is trapped by the glass and can be used for space heating or for heating water. This trapping of heat is called the Greenhouse Effect.

Figure 1

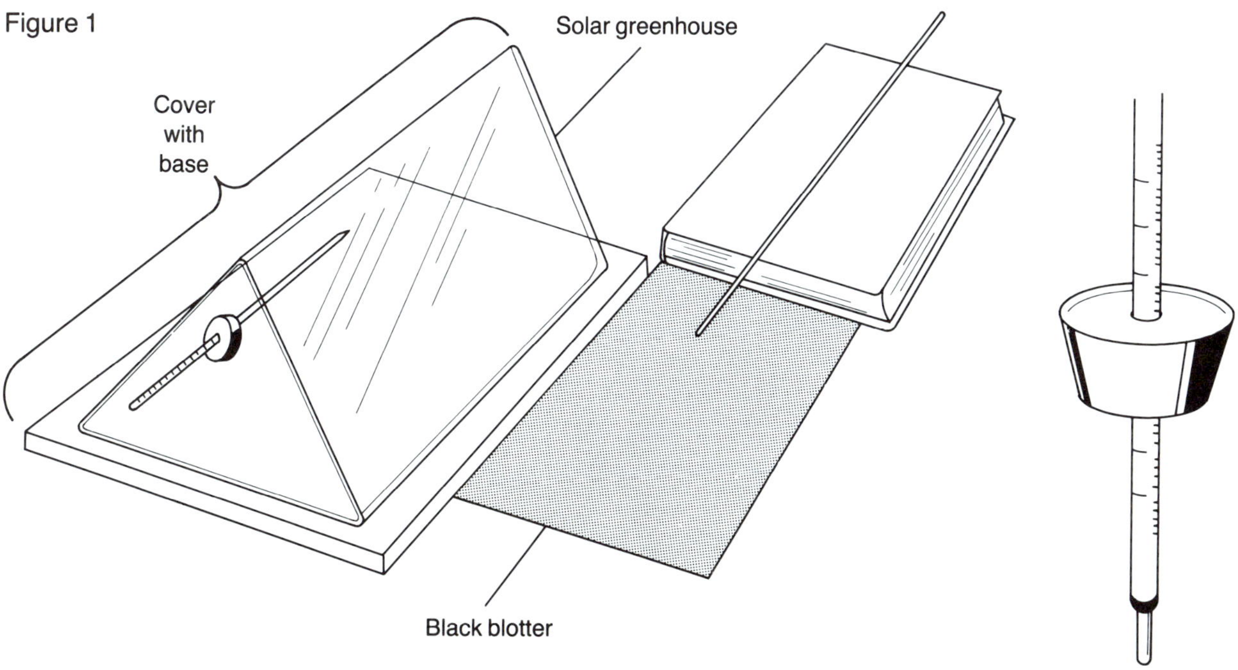

Name ______________________ Class ______________ Date ______________

Procedure:

1. Place a thermometer on a book (or other object) about 5 cm above a piece of white paper.

2. Place a thermometer on a book or other object about 5 cm above a piece of black paper.

3. Obtain a black base and plastic cover. Put the plastic cover over the base, and then place a thermometer (with rubber stopper) in the hole drilled in one end of the cover. See Figure 1.

4. On a separate sheet, record the temperature of each thermometer.

5. Place all 3 setups in direct sunlight for 15 minutes. Then record the temperatures on the same sheet as the other data.

6. Answer the following questions.

a. Which thermometer had the highest temperature?

The thermometer in the solar collector.

b. Why was this temperature higher?

Glass or plastic enclosure traps the heat.

c. Give examples of other situations where you have observed the Greenhouse Effect.

In greenhouses, cars parked in the sun, and rooms with south-facing windows.

d. Study solar collectors in use in your neighborhood and describe how they work.

(Various answers possible)

Report format: Hand in the data sheet and answers to questions.

ACTIVITY 14

Some Properties of Soils

Purpose: To analyze and classify soil types, and to make some predictions about the behavior of soils

This exercise is most valuable if students collect the soil samples from known locations and relate them to the plant and animal populations in these areas.

Time required: 3 or 4 periods

Part 1. Analysis of Particle Size

Materials

soil samples including sand, loam, clay and silt soils
large beakers—1 per sample
magnetic stirrer
calgon dispersing solution
1000 mL graduated cylinders, 1 per sample
distilled water
U.S. Standard soil sieves, 1 set per group if possible
balance
vials with lids, 1 per sample
stirring rods
stopwatch or clock with second hand

Background information: Soil texture refers to the granular composition of soil. Ultimately, soil classification is based on the relative proportion of small particles (less than 2 millimeters in size). If particles greater than 2 millimeters are present in significant amounts, an appropriate adjective such as gravelly or stony is added to the name.

Soil particles are classified by the United States Department of Agriculture as follows:

very coarse sand (2.0–1.0 mm)
coarse sand (1.0–0.5 mm)
medium sand (.5–.25 mm)
fine sand (0.25–0.10 mm)
very fine sand (0.10–0.05 mm)
silt (0.05–0.0002 mm)
clay (< 0.0002 mm)

Sand size particles feel gritty when the soil is rubbed between the fingers. They are not plastic or sticky when wet. Silt particles feel smooth and powdery when rubbed between the fingers. They are not plastic or sticky when wet. Clay particles feel smooth, sticky and plastic when wet. They form very hard clods when dry.

Soils can be divided into three general textural classes of coarse-, medium- and fine-textured soils. These major classes are further broken down to give textural classes, which are described in Part 2. Texture is an extremely important characteristic of the soil. It affects the chemical, physical and biological properties of the soil. Fine-textured (predominance of clay) soils, for example, have a greater amount of surface area than coarse-textured (predominance of sand) soils.

A gram of loam soil exposes 75 to 195 square meters of surface area. A single gram of clay loam soil contains 5 to 6 acres of surface area.

Name ______________________ Class ______________ Date ______________

Procedure:

A. Settling time

1. Weigh out 100 grams of soil. Half fill a large beaker with distilled water and add 10 mL of calgon dispersing solution.

2. Put the beaker on the magnetic mixer, add a large stirring magnet and turn on mixer so the water is stirring vigorously. Slowly pour soil into the beaker and mix for 10 minutes.

3. Pour the mixture into a 1000-mL graduated cylinder. Be sure to empty all the soil into the cylinder. Add enough distilled water to bring the water level up to 1000 mL.

4. Stir the contents vigorously for about 1 minute using a stirring rod. Stop stirring and record the time. At exactly 40 seconds, measure the amount of soil that has settled on the bottom of the graduated cylinder. This is the *sand* portion; record it on a data sheet in mL or mm.

5. Do not disturb the cylinder for 2 hours. At the end of exactly 2 hours, measure the total amount of soil which has settled to the bottom of the cylinder. This is the amount of sand and silt. Subtract the amount of sand and record the *silt* portion on the same data sheet in mL or mm.

B. Dry particle size

1. Weigh out 100 grams of dry soil.

2. Prepare a set of sieves in either of the size sequences listed below:

Table 1

5	______	5	______	
7	______	7	______	
9	______	10	______	Soil source
24	______	25	______	______________
48	______	50	______	
60	______	60	______	
250	______	270	______	
pan	______	pan	______	
Total	______	Total	______	

3. Pour the dry soil into the sieve series, starting with the largest one. Shake the sieves to make the soil settle to the smallest sieve possible. Break up any larger particles as needed.

4. Lay out a series of paper towels and label each with the mesh size of one of the screens.

5. Pour soil from each sieve on appropriate towels.

6. Weigh each size sample and record on Table 1. When totalled, it should add up to 100 grams.

C. Wet particle size

1. Obtain a vial with lid to match the drawing to the right.

2. Fill the vial to half with soil.

3. Add distilled water to fill to the neck of the vial.

4. Seal tightly and shake vigorously for 1 minute. Add water to neck and repeat.

5. Allow the vial to stand for 1 hour.

6. Sketch the contents and label them as best you can.

Figure 1

Part 2. Soil Texture by Feel

Materials

soil samples	water

Background information: Basically, soils fall into 3 major groups of soils. Those that form good ribbons (when they are wetted and shaped thusly) because they contain a large proportion of clay are called clays; soils that form a medium ribbon are called clay loams; and soils that form poor or weak ribbons are called loams. Sands do not form ribbons.

The following guide indicates the feel of each soil textural class:

Sand: Soil is loose and single-grained. The individual grains can readily be seen or felt. If squeezed in the hand when dry, sand will fall apart when the pressure is released. Squeezed when moist, it will form a cast, but will crumble when touched.

Loamy sand: Soil has characteristics of sand, but will make casts that are slightly more stable.

Sandy loam: Soil contains much sand but has enough silt and clay to make it somewhat coherent. The individual sand grains can be readily seen and felt giving a gritty feel. Squeezed when dry, it will form a cast which will readily fall apart, but if squeezed when moist a cast can be formed that will bear careful handling without breaking.

Loam: Soil has a relatively even mixture of the different grades of silt and clay. It has a somewhat gritty feel, yet is fairly smooth and slightly plastic. Squeezed when dry, it will form a cast that will bear careful handling, while the cast formed by squeezing the moist soil can be handled quite freely without breaking. Clods are soft and easily broken.

Silt loam: Soil contains a moderate amount of the fine grades of sand and only a small amount of clay, over half of the particles being of the size called silt. When dry it may appear quite cloddy but the lumps can be readily broken, and when pulverized it feels soft and floury. When wet the soil readily runs together and puddles. Either dry or moist it will form casts that can be freely handled without breaking, but when moistened and squeezed between thumb and finger it will not ribbon but will give a broken appearance.

Clay loam: Soil is fine-textured, usually breaking into clods or lumps that are hard when dry. When the moist soil is pinched between the thumb and finger it will form a thin ribbon which will break readily, barely sustaining its own weight. The moist soil is plastic and will form a cast that will bear much handling. When kneaded in the hand it does not crumble readily but tends to work into a heavy compact mass.

Sandy clay loam: Soil is similar to clay loam but is distinctly gritty.

Silty clay loam: Soil is similar to clay loam but is velvety when moist.

Clay: Soil is fine-textured, usually forming very hard clods when dry and is quite plastic and usually sticky when wet. When the moist soil is pinched out between the thumb and fingers, it will form a long flexible ribbon.

Sandy clay: Soil is similar to clay but somewhat gritty.

Silty clay: Soil is similar to clay but velvety when moist.

Name ______________________ Class ____________ Date ____________

Procedure:

1. Using the soils provided, determine whether the soil best fits into a clay, clay loam, loam, or sand class by determining whether it makes a good, medium, poor, or no ribbon as described below.

2. Place about ½ to 1 teaspoon of soil in the palm of your hand.

3. Add water very slowly, drop by drop, from a water bottle or tap.

4. Knead the soil all the while and bring it to the consistency of moist, workable putty.

5. When the soil is at the proper consistency, try to press it into a ribbon between the thumb and forefinger.

6. Compare the ribbon and the feel with the guides given under "Background information."

7. Record your results as shown below.

Table 2

Soil number	Class
1	____________
2	____________
3	____________
4	____________

Name ______________________ Class ____________ Date ____________

Part 3. Soil Percolation Rate

Materials

soil samples
ring stand
1″ diameter plastic tubing, at least 5″ long
small beakers
cheese cloth
rubber bands

Background information: The soil percolation rate measures how fast water moves through the soil sample. Soils with a low percolation rate retain water longer than soils with a high percolation rate. You will determine the rate by measuring how much water is left in the soil sample after a given amount of time has passed.

Procedure:

1. Prepare a ring stand as shown at the right.

2. Add 100 mm of soil, packing it lightly to make up the 100 mm.

3. Add 100 mL of tap water and note the time. Stop the timing after 45 minutes. Collect the water in a small beaker under the tubing.

4. Measure the amount of water collected in the beaker and subtract from the original 100 mL. This is the amount of water retained in the soil.

5. Record your results in a table as shown below.

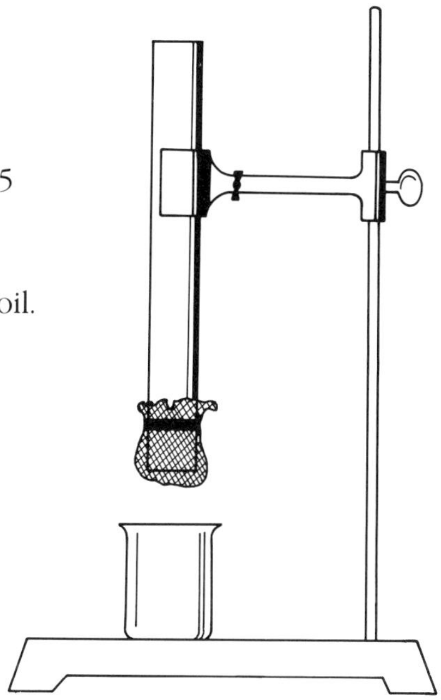

Figure 2

Table 3

Soil number	Percolation time	Amount of water retained in soil
1	45 min.	________
2	45 min.	________
3	45 min.	________
4	45 min.	________

Part 4. Discussion and Summary of Results

1. Based upon the data you collected, classify each soil:

Table 4

Class	Location
1. ________	______________________
2. ________	______________________
3. ________	______________________
4. ________	______________________

Name ______________________ Class ______________ Date ______________

2. What is the relationship between particle size and percolation rate?

The larger the particle size, the greater the percolation rate.

3. What does particle size have to do with retention of water in the soil?

The larger the particle size, the lower the water retention.

4. Which soil had the most even distribution of particle sizes?

(Depends on type of soils brought into class.)

5. Which soil had the most uneven distribution of particle sizes?

(Same as 4.)

6. What size particle sinks in water most slowly?

Smallest particles like silt and clay sink the slowest.

7. What size particle would travel farthest in moving water such as a river?

Silt.

8. On a steep slope, what size particle would wash away most easily?

Silt.

9. As a result, how would you describe mountain soils based on particle size?

Mountain soils are often made of the largest particles.

10. Beach sand in a given area tends to be composed of particles very nearly equal in size. Suggest a reason for this.

It's eroded from inland regions and would be deposited uniformly.

Name ______________________ Class ____________ Date ____________

11. In Upper Newport Bay, flood waters containing soil flow rapidly into the upper part of the bay, then slow down as they spread out over a large flat expanse of water area. Farther down the bay, these flood waters move very slowly into channels where there are homes and boat slips. Describe the likely particle size of the material on the bottom of these channels.

Large particles would deposit first in the bay. In the channels silt would probably be found.

Report format: Hand in data sheets and answers to questions.

A C T I V I T Y 15

Learning About Energy Quality

Purpose: To determine whether wood or paraffin has more energy per unit weight

Time required: 1 period

Materials

small tin can or small beaker
ring stand and wire triangle
thermometer
2 crucibles or foil cups
small pieces of wood and paraffin
balance or postage scale
graduated cylinder

Name ______________________ Class ____________ Date ____________

Procedure:

1. Weigh out 30 grams of wood splinters or small sticks and place in a crucible.

2. Measure 50 mL of room-temperature water and place in the small beaker. Record the temperature of the water.

3. Place the wire triangle on the ring stand with the crucible under it and the beaker on top. Adjust the height so that there is about 1″ between the top of the crucible and the bottom of the beaker.

4. Ignite the wood and allow it to burn completely. Immediately record the temperature of the water.

5. Repeat the procedure using 30 grams of paraffin. Be sure to record the water temperature before and after heating.

Use caution with melted paraffin. Birthday candles burn easily; disregard the weight of the wicks in making your calculations.

6. Answer the following questions:

a. Which has more energy, wood or paraffin?

Paraffin has more energy per unit mass than wood.

b. The calorie is a unit of energy. How many calories are contained in 30 grams of each material? To calculate this we assume that all the calories went into heating the water, and use this formula:

1 calorie heats 1 gram of water 1 degree Celsius, by definition, and 1 milliliter of water weighs 1 gram.

(Number of grams × rise in temperature in degrees Celsius)

Name ______________________ Class ____________ Date ____________

c. Would it be more energy efficient to heat your house with wood or paraffin, considering the energy needed to transport both these materials?

____

d. Wood is a renewable resource, while paraffin is derived from petroleum and is nonrenewable. Decide which fuel would be more desirable for space heating. Does this depend on where you live?

____

Report format: Hand in your data, calculations and answers.

ACTIVITY 16

Air Pollution

Part 1. Observation of Particulate Air Pollution

Purpose: To measure the particles in the air, and to determine whether certain locations have more particles than other locations

Time required: 1 period to set up the petri dishes, then 3 days' wait, then 1 period to count the particles and complete the data table

Materials (for each group of students)

12 petri dishes	graph paper
small amount of petroleum jelly	magnifying glass

Figure 1

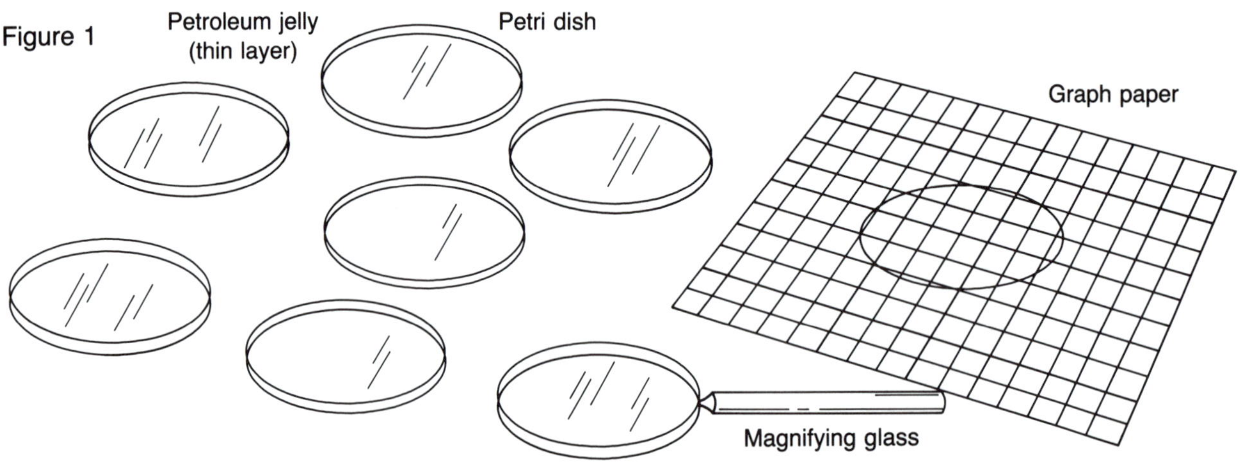

Procedure: Write your answers on a separate sheet.

1. Spread a thin layer of petroleum jelly on the flat surface of each of the 12 petri dishes.

2. Leave 2 dishes in each of 6 different, safe locations where they will not be disturbed for 3 days. Include areas that you suspect have high, low and average levels of air pollution.

3. Trace around the base of a petri dish on graph paper. The squares of the graph paper form a grid, which will aid you in counting particles.

4. After collecting the dishes, place them one at a time on the graph paper circle. Using a magnifying glass, count the number of particles inside each square of the graph paper and total them for each dish. Total the number of particles recorded for each pair of dishes from the same

You may wish to assign only one location to each group. Coal-fired power plants and incinerators produce many particulates; one location may be near one of these. You may want to have students measure indoor air in rural and urban locations or in households where smokers reside. Help the students practice counting so that their results will be consistent.

location and divide by 2 to get the average number of particles per location. Record your results on the data sheet below.

5. Compare the averages from different locations and explain why you think they differ. Also compare your averages with the averages from other groups and suggest reasons for any differences you find.

Name ______________________ Class ____________ Date ____________

Data Sheet

Dish #	Location	# of particles	Average # of particles
1			
2			
3			
4			
5			
6			
7			
8			
9			
10			
11			
12			

Report format: Hand in your data sheet and answers to question 5 above. Suggest possible sources for the particles you observed.

Part 2. Sources of Air Pollution

Purpose: To relate types of pollutants to the types of energy sources utilized

Time required: 1 period to organize the project, and at least 1 more period to discuss the results after the information has been collected

Materials

paper
pencil
telephone

Procedure: Write your answers on a separate sheet.

Accurate information and unbiased viewpoints on this issue may be difficult to find. Students should be encouraged to think of personal solutions to help reduce air pollution.

1. Contact local or state agencies and collect information on the types and amounts of air pollution in your area. Determine the pollution sources (automobiles, factories, airports, etc.) in your area. The Environmental Protection Agency (EPA) should have this information. Be as specific as possible.

2. Select 1 type of air pollution and design a plan that would reduce or eliminate that type of pollution. Try to be both imaginative and practical at the same time.

3. You may also want to find out about some careers related to air pollution and the training needed for these.

Report format: Prepare a list of pollutants found in your area and their sources, plus a short report of your ideas on reducing pollution.

ACTIVITY

Wet-Scrubber

17

Purpose: To become familiar with a wet-scrubber by building and using one

Time required: 1 or 2 periods

Materials

paper towels
12-cm piece of glass tubing
3 2.5-cm pieces of glass tubing
3 500-mL flasks
2 glass impingers (glass tubing drawn at one end to give it a smaller diameter so as to let out smaller bubbles)
heat source (burner or hot plate)
3 2-hole rubber stoppers (of a size to fit the mouths of the flasks)
2 30-cm pieces of rubber tubing
ringstand apparatus
vacuum source

Background information: The wet-scrubber is one of the most common pollution control devices used by industry. It operates on a very simple principle: a polluted gas stream is brought into contact with water so that the pollutants can be absorbed by the water.

Figure 1

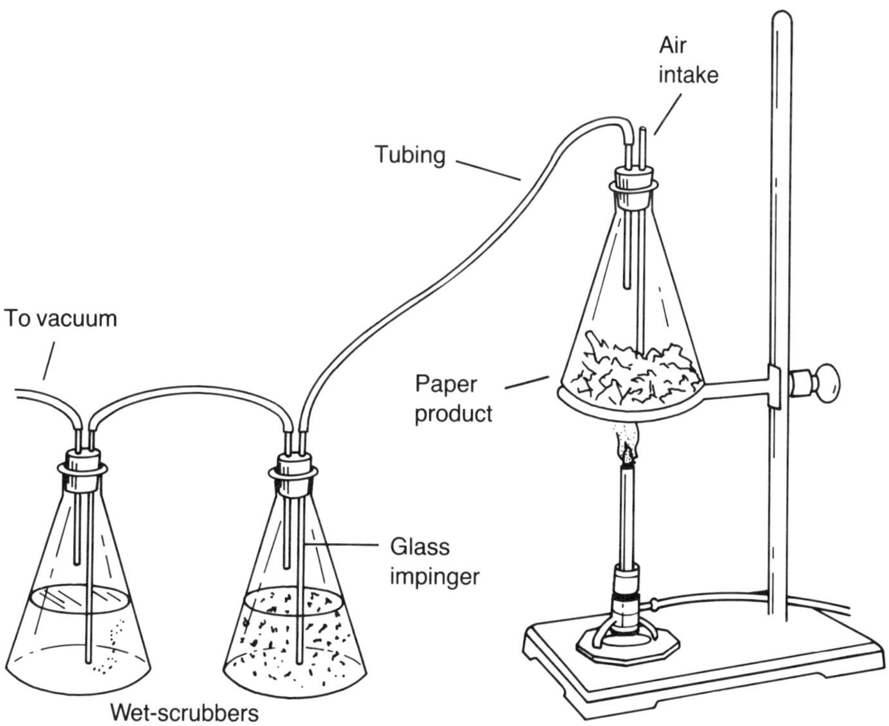

Name ______________________ Class ____________ Date ____________

Procedure:

Show students how to insert glass tubing safely. Remind students to remove the burner if too much smoke is being produced. Only a small amount of crumpled paper towel is needed, as shown in Figure 1.

1. Set up the apparatus as shown in Figure 1. Put a paper towel in a 500-mL flask and place this above the burner.

2. Using a 2-hole stopper that makes an air-tight seal with the flask, insert a 12-cm section of glass tubing through one of the holes. The tubing should reach to approximately 1.2 cm from the bottom of the flask.

3. Insert a 2.5-cm piece of glass tubing into the other hole of the stopper.

4. Connect a 30-cm piece of rubber tubing to the 2.5-cm piece of glass tubing, making sure an air-tight seal exists.

5. Fill a second 500-mL flask approximately ¾ full of water. Using a second double-hole stopper, put a 2.5-cm piece of glass tubing into one of the holes, and insert the glass impinger into the other.

6. Construct a third flask like the second. Use Figure 1 to check your apparatus.

7. Connect the rubber tubing as shown and heat the first flask (combustion chamber) until smoke appears.

8. Put a vacuum on the third flask to draw a stream of smoke through the second flask (the wet-scrubber). If smoke collects in the second flask above the water, a second scrubber can be added.

9. After observing the wet-scrubber, answer the following questions:

a. Why does the water in the wet-scrubber change color?

It captures pollutants, especially particulates given off by combustion.

b. Why does the wet-scrubber have an impinger (in other words, why is it important for small bubbles to be formed)?

To help pollutant gases dissolve in the water.

c. What does the scrubber filter out of the air? not filter out?

Effectively removes particulates and water soluble gases such as carbon dioxide and sulfur dioxide. Nitrogen dioxide is not very water soluble so it tends to escape.

Name ______________________ Class ____________ Date ____________

d. Suggest ways to dispose of the pollutants which are now trapped in the water.

Solid particles could be filtered out or allowed to settle out of solution. They could be disposed of in landfills with thick clay liners that prevent leakage. Sulfur dioxide in water converts to sulfuric acid, which could be chemically reacted with magnesium hydroxide, forming magnesium sulfate, which can be used to make dry wall.

Report format: Answer questions as shown.

ACTIVITY 18

Temperature Inversions and Air Pollution

Purpose: To model temperature inversions and (in Part 2) to demonstrate how temperature inversions trap air pollutants

Time required: 1 period for each part. Part 2 may be done as a demonstration.

Background information: The concentration of air pollutants in any region depends not only on the sources but also upon the region's weather. Wind speed and direction, together with the vertical temperature distribution, are the principal elements affecting the concentration.

For instance, the higher the elevation, the lower the average temperature of the air. A wind condition called *vertical turbulence* carries pollutants into regions of high elevation. Whenever the temperature at high elevations increases instead of decreases, vertical movements of air masses are restricted. A layer of cool air is trapped near the surface of the ground by a warmer air mass above it. This condition of restricted air movement is known as *temperature inversion,* one which traps pollutants in the area where they were produced.

Part 1. A Model Temperature Inversion Using Water

Materials

ice cubes
large plastic bag
aquarium (20–40 liters)
food coloring
hot and cold water
string or rubber band

Procedure:

1. Fill the aquarium ¾ full of cold water and add ice cubes to it. Allow the water time to get cold.

2. Fill a plastic bag with hot water. CAREFUL, not too hot!

3. Place enough food coloring in the water in the bag to color it. Seal the plastic bag with some string or a rubber band.

4. Remove all ice cubes from the aquarium and place the plastic bag near the top of the aquarium water.

5. When the water becomes calm, carefully pierce the bag with a pin in several places. Observe what happens to the hot water, then complete the following exercises.

Don't attempt to move full tanks. If possible set the tanks up near sinks and fill and empty them using hoses. Remind students that temperature decreases with altitude at a regular rate, even though warm air rises.

Name ______________________ Class ______________ Date ______________

a. Describe what happened to the hot and cold water in the tank.

b. Compare your observations here to what occurs in a temperature inversion.

6. You may want to reverse the positions of the hot and cold water—hot water in tank, cold water in bag—to see what happens.

Report format: Answer questions as shown.

Part 2. A Temperature Inversion Traps Pollutants

Materials

- 1-liter graduated cylinder
- 2- or 3-liter beakers
- mixture of ice, salt, and water
- 2 thermometers
- string or thread
- heat lamp
- ring stands and clamps
- glass tubing with elbow joint
- rubber tubing
- hand vacuum assembly (Millipore has such an assembly) (air piston or large syringe)
- cigarette

Figure 1

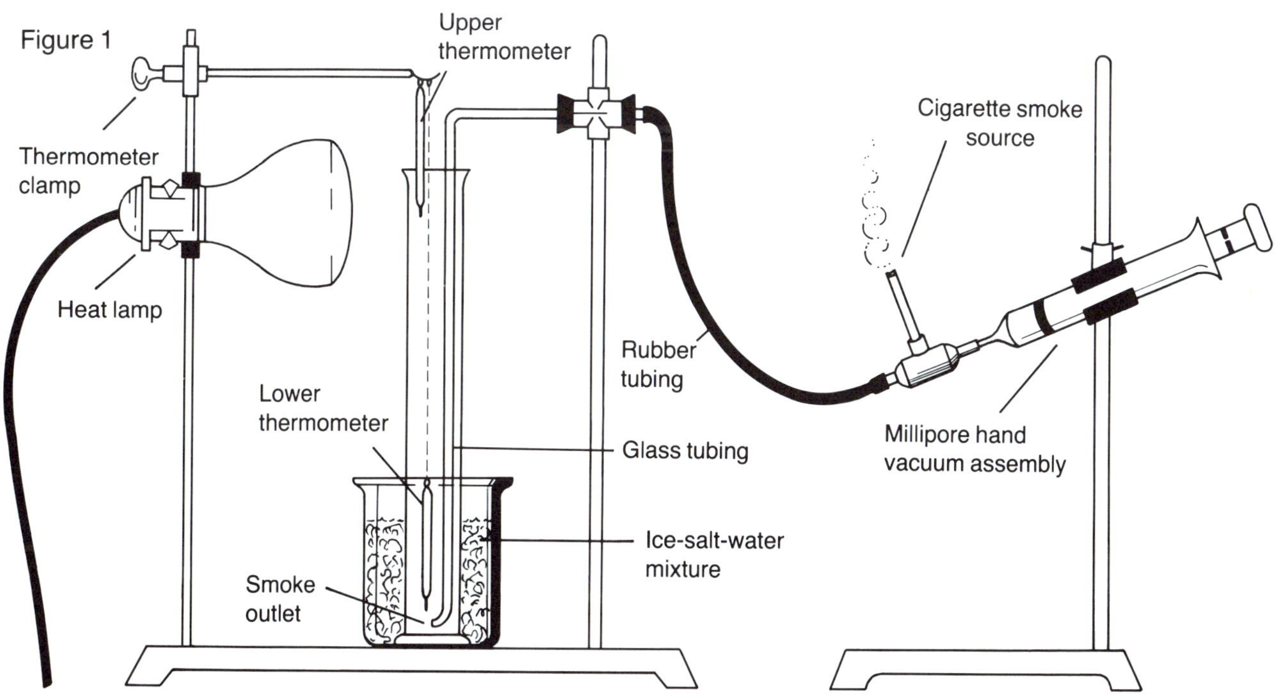

Name ______________________ Class ____________ Date ____________

Procedure:

Have students set up apparatus in an area with good ventilation.

1. Set up apparatus as shown in Figure 1.

2. Do not activate the smoke source or the heat lamp until ice-salt-water mixture has cooled the air in the lower portion of the graduated cylinder.

3. Use the heat lamp to warm the air in the upper portion of the graduated cylinder. When a significant temperature difference exists between the upper and lower regions of the cylinder, light the cigarette.

4. Slowly introduce the smoke into the bottom of the cylinder.

5. Observe the distribution of smoke within the cylinder.

6. After a dense layer of smoke has formed inside the cylinder, carefully remove the beaker containing the ice-salt-water mixture.

7. Note how the temperature changes affect the smoke dispersal. (To speed this process, move the heat lamp closer to the base of the graduated cylinder.) Demonstrate the dispersal of smoke under normal conditions; use the heat lamp to warm the air in the bottom of the graduated cylinder to a higher temperature than that of the air in the upper portion.

Questions:

a. What happens to the cigarette smoke during the temperature inversion (when cold air is on the bottom)?

Smoke remains trapped in the cooler, denser air.

b. What happens to the smoke when the inversion is destroyed by removing the ice-salt mixture?

Smoke rises in the column.

c. What combination of weather and geographic conditions must be present in order for temperature inversion to occur naturally?

Mountainous terrain and valleys trap pollutants in the cool ground air; cold weather in the winter often produces temperature inversions (see text for explanation of radiation inversions); high-pressure systems (see text for explanation of subsidence inversions).

d. Does temperature inversion occur in your area? If so, at what season of the year are they most serious?

(Various answers)

Report format: Answer questions as shown.

ACTIVITY 19

Nuclear Energy Viewpoints

Purpose: To consider the pros and cons of the use of nuclear power

Time required: 1 period to organize the activity; 1 or more periods for speakers and for discussion after information has been collected

Materials

paper	telephone
pencil	tape recorder optional

Procedure:

Consider showing the film "The China Syndrome" or other films or filmstrips dealing with nuclear energy.

1. Locate and study books on nuclear energy to get as much background information as possible. Also collect newspaper and magazine articles and letters on nuclear power. Find out as much as you can about the authors, and evaluate the accuracy and possible biases of the information.

2. Develop a speaker's list of qualified individuals, noting why you think they are qualified to speak about nuclear power.

3. Select several speakers, some in favor of nuclear power and some opposed. Tape record their presentations if they agree. Allow time for questions and answers at the end.

4. After all presentations, hold a general classroom discussion on nuclear energy. List pro and con arguments on the board, and use all information you have located. The discussion may take the form of a debate, with students assigned to pro and con positions.

Report format: Write a short paper stating your own position on nuclear power. Use references, including speakers, to support your position.

ACTIVITY 20

Ecological Debate

Purpose: To consider how humans fit into the ecological world

Time required: 2 class periods, with time in between for debate teams to prepare

Materials

textbook

supporting material on relevant issues

Procedure:

1. Select 2 5-member debate teams. Team A will argue the case for humans being apart from nature and here to dominate over other living things. Team B will take the opposite case—that humans are a part of nature and must learn to live in harmony. One member of each team should be selected as a leader.

2. Each group should meet outside of class to discuss its debate strategy. A suggested approach would be to draw up a list of arguments that support your case. You can use the text to find some of the supporting material, since it covers both sides of many issues. You may want to do further library research or talk to experts in the field to strengthen your case. Each group should also try to anticipate the arguments made by the opposing team. List what you think the opposition will say, and then list ways to refute their arguments.

3. In class, after both teams have prepared adequately, open up the debate. The debate referee will start the debate with questions he or she has drawn up. The referee will ask a question and then allow one member of each team to respond, keeping all responses down to one minute or some other manageable period. Class members can participate by writing questions out on note cards and submitting them to the referee prior to the debate. This way everyone can get involved.

4. Class members should take notes during the debate, noting the question, the response of team A and team B, and personal comments. Use the space on page 75 for your notes.

Name ____________________ Class ____________ Date ____________

Question	Response of Team A	 Team B	Personal Comments

5. When the debate is finished, the teacher should open the floor for comments. Which team presented the best case? Why? Which arguments were the most convincing? Which were the least convincing? What new insights have you gained?

Report format: Answer question as shown.

ACTIVITY 21

Popularizing Environmentally Favorable Lifestyles

Purpose: Devise ways to make environmentally favorable lifestyles more popular

Time required: 2 or more periods

Materials

art materials may be helpful

Background information: Many people realize that production of the goods and services they use damages the environment and depletes natural resources. Despite this, people are reluctant to change their lifestyles.

Madison Avenue in New York City has traditionally been known as the home of many of the world's top salespeople. Companies located there have popularized and sold attitudes which have affected and changed our eating habits, our clothing styles, and the kinds of entertainment we enjoy.

For this activity assume you are a member of one of these high-power Madison Avenue advertising firms. Your firm has been hired to popularize some changes in lifestyles.

Procedure:

1. Make up a team of 3 or 4 people. Your team will be assigned 1 of the "new habits" listed below. The team will then represent an advertising company which is competing with other advertising companies for a contract.

2. When you have developed a method that works, write a paragraph that describes how the activity could be made popular so that the average citizen begins doing it and, better yet, wants to do it.

3. Arrangements will be made by an impartial panel that will decide which three advertising firms get an account.

4. Here are the new habits you must sell:

a. Judicious or conservative use of energy. Each home and person uses energy efficiently for heating, cooking, refrigeration and cleaning.

b. Use shower and bath water conservatively, except in cases of illness. Water is to be considered as precious as gold.

c. Purchase sturdy clothing to satisfy basic needs, repair them yourself when they become damaged, and buy new clothing only when necessary.

d. Eliminate the use of pesticides and chemicals around the house.

e. Avoid junk food. Buy and eat only the nutritious foods you need. The average adult human requires about 2,500 calories a day for normal activity.

f. Eat lower on the food chain using substitutes for meat protein, such as cheese.

g. Popularize environment-conscious forms of recreation such as sailing, bicycling, and canoeing in order to reduce consumption of fossil fuels.

h. Reduce population growth.

i. Greatly reduce garbage service and increase recycling and reuse. All organic waste is composted.

j. Use public transportation whenever possible.

This activity is designed to help students focus on some of the positive aspects of a conserving lifestyle. After the presentations, encourage students to consider smaller changes which might actually be made.

Report format: Each team will give a presentation to the rest of the class. The panel will rate each team on persuasiveness.

ACTIVITY 22

Biodegradability of Solids in a Mini-Landfill

Purpose: 1) To determine the rates of decomposition of various substances in a model landfill, and 2) to consider the effectiveness of landfills for waste disposal

Time required: 1 period to set up, then a few minutes weekly for 4 weeks

Materials

1 clear plastic drinking glass
aluminum foil square to cover glass
solid wastes—glass, paper, metal, fruit, bread, plastic, wood, etc.
garden or potting soil

Background information: As both the population and the amount of trash per person increase, new methods of disposing of garbage are being sought. Landfills are used in many areas. Typically, garbage is dumped into a large depression in the earth and covered with soil. This process may harm the soil, as landfill soil contains less air than undisturbed soil. This may slow down decay rates. Some types of wastes are more suitable for disposal in this way than others.

Procedure:

1. Make a mini-landfill using a clear plastic glass as a container. Fill the glass with garden soil and bury pieces of different solid wastes at various levels next to the inner surfaces of the glass. Set these up so that you can see each piece. You may want to measure or weigh each piece before burying it. Leave 1″ of air space over the top of the soil and pack firmly. Keep the soil moist but not waterlogged. If you want to study the effect of water on decomposition, set up a second glass and keep it dry. Cover the glass with aluminum foil.

2. Observe any changes in the materials over a 4-week period. Record your observations on the following chart.

Name ______________________ Class ______________ Date ______________

Table 1

Sample	Beginning date:		After week 1 date:		After week 2 date:	
	Measurement	Observation	Measurement	Observation	Measurement	Observation

Sample	Beginning date:		After week 3 date:		After week 4 date:	
	Measurement	Observation	Measurement	Observation	Measurement	Observation

3. After 4 or more weeks, answer the following questions:

a. What type of material changed the most after 1 week?

Fruit and bread.

b. What type of material was just beginning to change after 4 weeks?

Paper.

c. Which material did not change at all?

Glass, plastic and metal.

d. Which materials do you think might not disappear from a landfill for 100 years or more?

Glass, plastic and metal.

Name __________ Class __________ Date __________

4. Suggest answers to the questions below based on your observations:

a. What are some problems that might arise when a landfill area is used for building homes?

Two of the chief problems are subsidence and methane gas emission, which can result in explosions.

b. Does moisture influence decomposition?

Moisture greatly accelerates decomposition.

c. What materials should not be put in a landfill?

Nonbiodegradable, recyclable materials that could be melted down and reused.

5. Find out if there are any large landfills in your area. If so, what are the future plans for the use of these areas?

Report format: Hand in data sheet and answer questions as shown.

ACTIVITY

Does Insulation Really Help to Conserve Heat?

23

Purpose: To determine how well different types of insulation reduce heat loss in a model system

Time required: 1 period

Materials

2 small cans or 100-mL beakers
1 large can or 500-mL beaker
watch or clock
aluminum foil
hot water
2 thermometers
graph paper
different types of insulation, including foam plastic bits, sawdust, cork, dry or wet sand, cellulose, fiberglass or clear bubble packing material

Figure 1

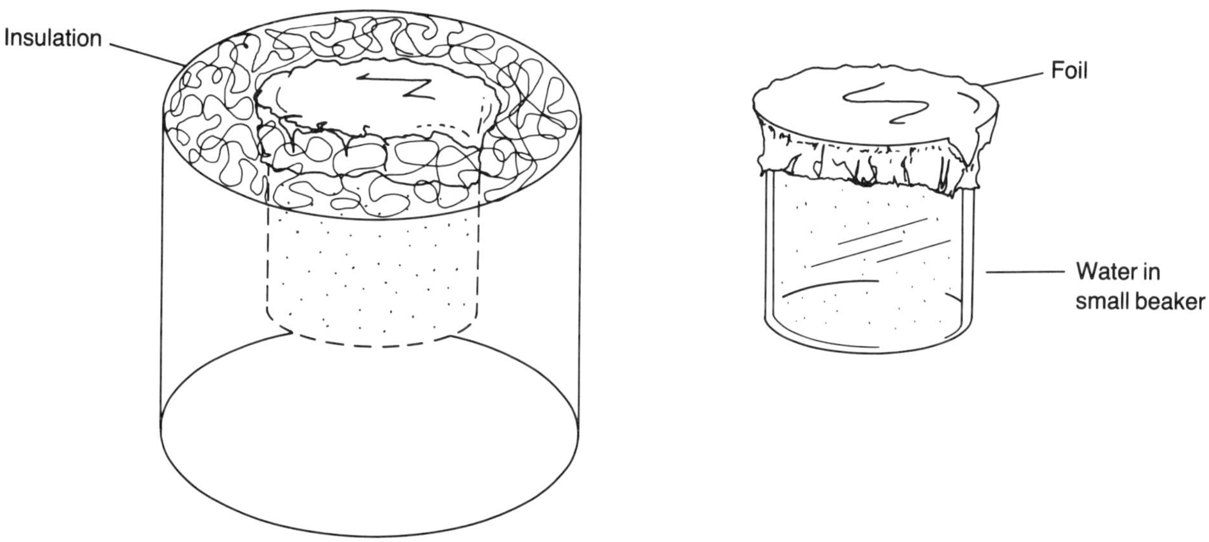

Procedure:

1. Decide which types of insulation you want to test. You will need 1 set of the materials listed above for each type.

2. Prepare a data sheet. Include a column for time, 1 for temperature of the uninsulated beaker, and 1 column for each of the insulated beakers you will be setting up.

3. Put a ½″ layer of insulation in the bottom of the large beaker. Place the small beaker on top of this layer and fill the space around it with insulation.

4. Fill both small beakers nearly to the top with hot water and cover them with aluminum foil (see Figure 1).

Name ______________________ Class ______________ Date ______________

5. Insert a thermometer through the foil cover of each beaker. You should be able to read the thermometer without removing it from the beaker. Record the temperatures and time on the log sheet.

6. Read the thermometers every 15 minutes and record the temperatures on the log sheet. Continue for 90 minutes, if possible, or until both systems reach the same temperature.

7. Graph time versus temperature for both systems.

8. If time allows, you may want to set up systems to test whether the thickness of the insulation influences heat loss. Would you get the same results by allowing cold water to warm up to room temperature? You could also try leaving the systems outdoors in winter, or in a refrigerator.

The thickness of the insulation reduces the heat loss. Insulation reduces the rate at which a beaker of cold water heats up to room temperature.

9. Compare your results with those of other class members and answer the following questions.

a. Which material was the best insulator? Which material was the least effective?

(Answers depend on materials used)

b. How do you know that insulation slows but does not stop heat flow?

Insulation slows heat loss but does not prevent it because the hot water will eventually reach room temperature.

c. Would the results of this experiment help you decide what kind of insulation would be best for your house? What other factors would also be important in making this decision?

Yes. How easily the insulation can be applied, its cost, potential health effects, and fire safety.

Report format: Hand in data sheet, graph and answers to questions.

ACTIVITY 24

Thermal Pollution

Purpose: To determine the effects of thermal pollution on goldfish and other fish

Time required: 1 period for Part 1; Part 2 may be done as homework

Figure 1

Background information: Many industrial processes produce hot water as a waste product. For example, electrical and nuclear power plants use large quantities of water from lakes and rivers to cool their equipment. The water becomes hot as it circulates around and cools the equipment. It is then often pumped back through a pipe into the river or lake from which it came.

The cooling water may be free of chemical pollutants. Nevertheless, because it is still hot, the temperature of the water near the power plant outlet can be raised as much as 4° to 12°C above the normal temperature. This is thermal pollution.

You may be wondering right now, "Wouldn't such large amounts of hot water affect animals and plants near the power plant outlet?" In this activity you will be given a chance to collect the data necessary to answer that question. Goldfish will be the experimental animal. Because temperatures are critical, follow the instructions carefully.

Part 1. The effects of temperature change on goldfish

Materials

- 2 containers (coffee cans or large beakers)
- alcohol burner or Bunsen burner, stand and wire screen
- timer (or watch with second hand)
- baby food jar and lid
- fish (goldfish)
- dip net
- thermometer
- dropper bottle H_2O_2 (medicine grade; purchased over the counter at a drugstore)
- distilled water

Procedure:

1. Make sure your containers are clean and free of soap.

2. Place about 300 mL of tap water in one container and 300 mL in the other container. Label A and B.

3. Fill the baby food jar to the top with distilled water or with water that has been standing overnight in an open container. Add 4 drops of H_2O_2 solution. (This will ensure a good oxygen supply for the fish.) Using a dip net, add 1 fish from the stock tank to the jar. Cap the jar and place it in container A.

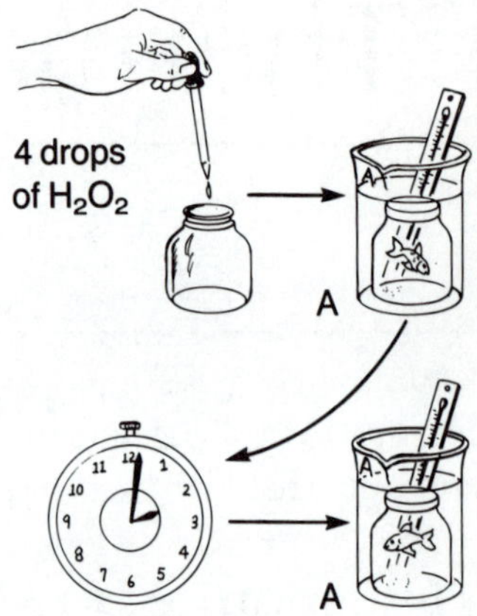

Figure 2

4. After a few minutes, record in Table 1 the temperature of the water in container A. Then determine the relative activity of the fish by counting the number of respiratory movements (opening and closing of the mouth) that occur in a 20-second span. Have your partner time you as you count. Take at least 3 complete 20-second counts to find the average. Record all measurements in Table 1. Be patient—your fish may move about, requiring you to start your count over again.

5. Put container B on the burner stand and heat the water. While the water is heating, add 2 more drops of H_2O_2 solution to the baby food jar still containing the fish. When the water temperature reaches 42°C, remove the burner and place the capped baby food jar in container B.

6. When the temperature of the water in container B has dropped to 38°C, determine the activity of the fish as you did before in step 4. Record in Table 1 this data and other observations you make of the behavior of the fish. Again, be patient in observing your fish. It may take a while to get 3 20-second counts.

7. When you have completed the activity, put the fish into the rest tank so that another student won't use it right away.

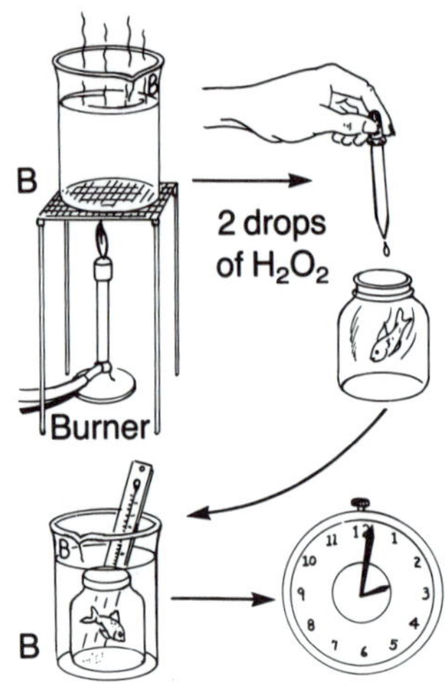

Figure 3

Interpretations:

TABLE 1

Container	Temperature	Respiratory Movements per 20 sec.	Average Count	Other Observations
A				
B				

1. When the fish was placed in the hot water, did its activity increase or decrease?

The fish's activity increased.
____

Name ______________________ Class ____________ Date ____________

2. What observations led you to your conclusion for the answer to 1?

Increased respiration and increased activity.

3. What evidence, if any, have you collected that would indicate that your fish was "uncomfortable" in the warm water?

Gasping for air at the surface.

Report format: Turn in data table and answer questions.

Part 2. The effects of thermal pollution on other fish

Background information: A fish is "cold-blooded," which means that its body temperature is close to the temperature of its environment. Temperature changes in the water lead to changes in the temperature of the blood circulating through the fish's gills. As the blood flows through the fish's body, the body temperature changes. With this in mind, answer the following questions.

1. Suppose an ocean sunfish had been swimming north in the warm waters of the Gulf Stream. When the fish entered the cold waters off the Maine coast, how would its temperature change?

Its body temperature would decrease.

2. Increasing the temperature speeds up the consumption of oxygen and the production of carbon dioxide by the fish. How does this information explain the change in breathing movements that you observed in your goldfish?

Requiring more oxygen, the goldfish increased its rate of respiration and went to the surface for more air.

3. If the water temperature were increased by another 5°C (from 38° to 43°C), what change in the fish's activity would you expect to observe?

Much faster rate of respiration, severe stress; fish could die.

4. What do you predict would happen to the breathing rate of the fish if the water were gradually cooled?

Rate of respiration would decrease.

Name ______________________ Class ______________ Date ______________

5. Besides goldfish, what reactions would you predict of other fish placed in warm water? Would they react as the goldfish did? Examine Table 2 below, which shows the thermal death point (temperature when death occurs) for various fishes.

Yes, but the cool-water fish, such as trout, would die more quickly.

6. Which fish are most sensitive to thermal pollution?

Trout and salmon.

7. Which fish are most tolerant of thermal pollution?

Goldfish, shad, bass, and bluegills.

Report format: Answer questions as shown.

TABLE 2

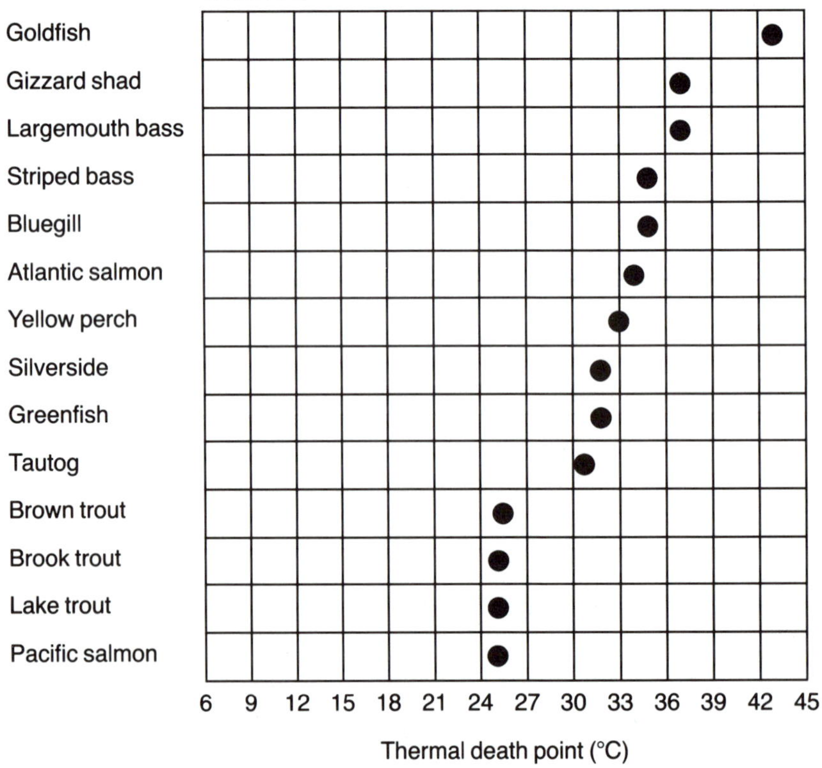

ACTIVITY 25

Investigating the Biological Community of a City

Purpose: To study a city as a biological community by identifying the plants and animals of the city, determining their relationships to each other and to human beings, and tracing the city's energy flow

Time required: 3 or more periods

Materials

binoculars

field guides to plants, birds, mammals, insects and possibly reptiles and amphibians

notebook to record observations

calculators may be useful

Name ____________________ Class __________ Date __________

Procedure:

1. Take a field trip through your neighborhood and identify as many plants and animals as you can.

2. Of the organisms you observed, which obtain their energy directly through photosynthesis? Also, list those which do *not* get their energy by photosynthesis; what are their energy sources?

____

____

____

What proportion of the biological energy needed by city organisms (including people) is produced inside the city, and what proportion comes from outside? On a separate sheet, construct a food web for the city.

____

3. Which organisms are present in the city only because human beings are present? In what way does their existence depend on human beings?

____

____

Name ______________________ Class ____________ Date ____________

4. Which organisms would exist in the area even if humans were not present?

Wild birds, rodents, insects, plant life.

Report format: Answer the questions above. Be sure to indicate what you observed, and what you estimated or assumed.

ACTIVITY

Science and the Scientific Method

26

Purpose: To practice gathering scientific information, testing ideas, and solving problems

Time required: 1 period

Materials

paper and pen

Background information: Science is a dynamic process—scientists identify problems, search for new information, and test their ideas through experimentation. Experimentation is the heart of the scientific method and the foundation upon which science rests. Refer to Chapter Supplement 1-1 (pages 26–28) to become familiar with the following terms: *fact, hypothesis, experiment, control, controlled experiment, experimental variable, data, analysis, conclusion.*

Procedure:

Read this fictional experiment and demonstrate the use of the scientific method by answering the questions that follow it.

Many people in the small midwestern town of Hootville are stricken with the disease Buggo. Most recover within seven to nine days.

Buggo has been shown to be caused by a bacterium called Gotcha. Antibiotic X is a new drug that kills Gotcha bacteria cultured in a test tube. Antibiotic Z cures dogs that have been stricken with Buggo.

Researchers decided to test Antibiotic X on some of the people in Hootville. They believed that the drug would effectively cure Buggo in humans, causing the patients who took it to recover more quickly than those who did not take it.

The researchers prepared tablets containing Antibiotic X. They also prepared tablets that did not contain Antibiotic X. A tablet that does not contain any medication is called a placebo.

Researchers selected two groups of people who had just contracted Buggo. Twenty-five people—Group A—were given tablets containing the drug. Twenty-five people—Group B—were given placebos.

Twenty people given the drug recovered within one day. The other five in Group A took from seven to nine days to recover.

One of the people given a placebo recovered within one day. The rest of Group B took from seven to nine days to recover.

Name ______________________ Class ____________ Date ____________

Questions:

1. List three facts known before the start of the experiment.

Buggo is caused by the bacterium Gotcha. Antibiotic X kills Gotcha bacteria cultured in a test tube. Most people with Gotcha take seven to nine days to recover.

2. State the researchers' hypothesis.

Antibiotic X will cure Buggo in humans.

3. How do Groups A and B differ?

In Group A, patients are given Antibiotic X. In Group B, patients are given a placebo.

4. In a controlled experiment, why is there only one experimental variable?

There is only one experimental variable so that the effect of the factor being tested can be observed.

5. Why did researchers first test Antibiotic X on a small group instead of giving it to everyone in town who had Buggo?

Researchers were not certain that Antibiotic X would work. It might have produced undesirable side effects.

6. What are the data in this experiment?

Twenty people in Group A, who received Antibiotic X, recovered within one day. Five people in Group A recovered within seven to nine days. One person in Group B, who received the placebo, recovered within one day. Twenty-four people in Group B recovered within seven to nine days.

Name ______________________ Class ____________ Date ____________

7. What is your analysis of the data?

Antibiotic X takes a day or less to cure most people (80%) who have Buggo.

8. What are your conclusions?

The data support the hypothesis.

Report format: Answer the questions above and then have a class discussion about the experiment in Hootville and whether the scientific method was used.

ACTIVITY 27

Understanding Population Growth

Purpose: To learn how the world's population grows and to explore how this growth affects the environment

Time required: 1 period

Materials

paper and pen

Background information: Exponential, or geometric, growth takes place when a population increases by a fixed percentage. Over a period of time, statistics for a population increasing geometrically will show that the ratio of any statistic to the preceding one is always the same. Biologists study how long it takes a population to double (an example of geometric growth) as a way to gauge the rate of population growth. When the various influences on population growth are favorable, the doubling time is shorter than when influences are adverse (see Chapter 6, pages 114–140).

Procedure:

1. Using the data in Table A, draw a world population curve by plotting the information on Figure 1.

Table A World Population

(Growth rate 1.7 percent per year)

Year	Population in Millions
1990	5,150
2000	6,100
2010	7,250
2020	8,600
2030	10,150
2040	12,000

Name ______________________ Class ____________ Date ____________

Figure 1 World Population Growth

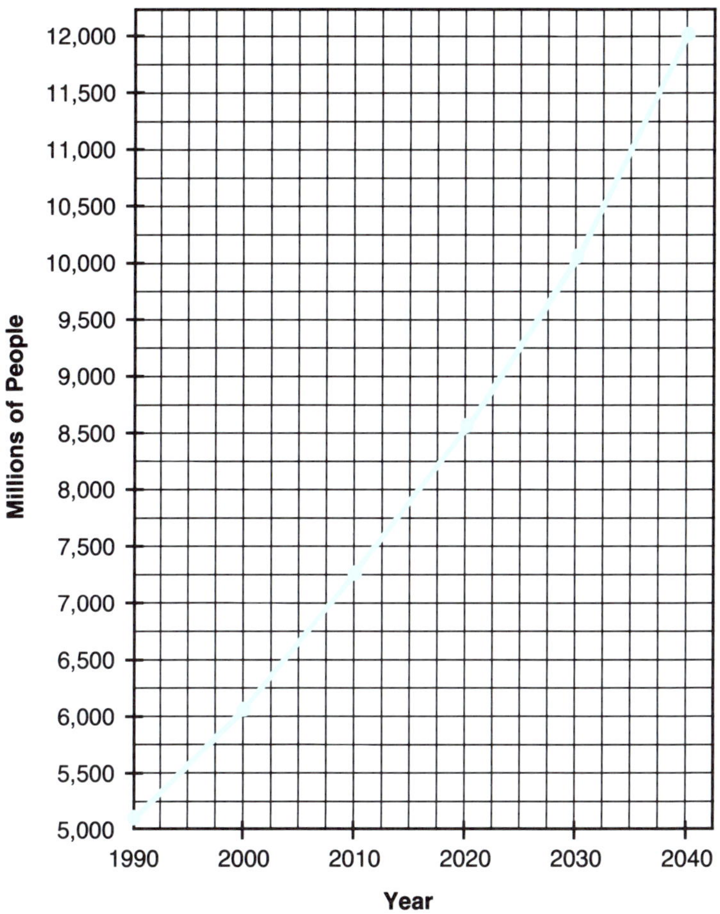

2. Refer to Figure 1 to answer the following questions:

a. Approximately how many millions of people will be in the world in the year 1995?

5,600. (Range from 5,500 to 5,700 is acceptable.)

b. What number is double your answer to 2a?

11,200 million. (Range from 11,000 to 11,400 million is acceptable.)

c. In which year will the population of the world reach the number you wrote down as the answer to 2b?

In about 2036. (Range from 2035 to 2037 is acceptable.)

Name ______________________ Class ______________ Date ______________

d. How many years will it take the world population of 1995 to double?

About 41 years. (About 40 years is acceptable.)

3. Starting at 1990, determine the approximate number of years it will take the world population to increase by 250 million (assuming a steady growth rate).

About 3 years.

4. Starting at 2035, determine the approximate number of years it will take the world population to increase by 250 million.

About 2 years.

5. In any population, certain environmental factors tend to limit the population's size. What environmental factors could affect the size of the world's human population?

Space, available food and fuel, disease.

Report format: Answer the questions above and turn in the graph.

ACTIVITY

Land Use

28

Purpose: To identify school and community open spaces and consider the consequences of developing them

Time required: 2 class periods; 1 hour outside class

Materials

paper and pen

automobile association map of your area

Name ____________ Class ________ Date ________

Procedure:

1. Walk around the perimeter of your campus. List open spaces, including gardens, lawns, football and baseball fields. Describe the different uses for these open spaces. Approximately what percent of the campus is open space?

[Various descriptions and percentages.]

2. From your list, choose an open space area large enough for a building or a parking lot. What would you construct on the site? What would be the trade-offs of developing the open space?

[Various answers.]

Name ______________________ Class __________ Date __________

3. On a map of your area, mark the location of your school with an X. Referring to the map scale, determine the length of a line representing 100 km. Using the X as the center, draw a circle with a 100-km radius. List the open spaces within the circle; include city parks, rangelands, and agricultural lands.

(Various answers.)

4. Choose one open space from your list and describe a possible way to develop the site. List some potential benefits and disadvantages of converting open space to this use. Considering the possible consequences for people, wildlife, and plants, do you think the change would be worthwhile?

(Various answers.)

Report format: Answer the questions.

ACTIVITY

The Effects of Toxics on Living Organisms

29

Purpose: To investigate the effect of a toxic substance on an aquatic organism

Time required: 1 period

Materials

8 petri dishes
felt-tip marker
10-mL graduated cylinder
80 mL of distilled water
medicine dropper
dilute copper sulfate solution
1 jar common culture *Daphnia* or live brine shrimp *(Artemia)*

Background information: Toxic substances discharged into bodies of water such as rivers, lakes, and streams have effects that are often not detected until they seriously damage the environment. Dead fish and birds are signs of water pollution, signaling that certain practices are unsafe.

Procedure:

1. Label the bottom of each of eight petri dishes: One is marked 0; two, 2; two, 4; two, 7; one, 10. Separate the dishes according to the numbers, which identify the number of drops of copper sulfate solution to be added.

2. Measure 10 mL of distilled water into a graduated cylinder. Pour the water into a petri dish. Repeat this step for the remaining seven dishes.

3. Pipette the specified number of copper sulfate solution drops into each of the eight dishes. (Remember, the dish labeled 0 will not have any copper sulfate solution. It is the control.) Cover each dish. To make certain that the chemical is dispersed uniformly in the water, swirl each dish first in one direction and then in the other.

4. Swirl the common culture of *Daphnia* to distribute the organisms uniformly throughout the culture. Doing this increases the probability that drops of the culture contain approximately equal numbers of *Daphnia* (be careful to dispense equal-sized drops). Add one drop of the culture to each of the dishes, including the control dish. To ensure continuing uniformity, swirl the common culture jar before dispersing a drop into each dish.

Name ____________________ Class ____________ Date ____________

5. After 5 minutes, carefully observe the *Daphnia,* then answer the following questions.

a. Where in each dish do the *Daphnia* congregate?

Daphnia will probably cluster near the middle of the dishes labeled 0, 2, and 4, and near the surface in dishes labeled 7 and 10.

b. In which dish are they moving most rapidly?

Dishes with lower concentrations of copper sulfate (0 and 2) should have the fastest-moving organisms.

c. How many drops of the copper sulfate solution affect *Daphnia* behavior without killing the organisms?

Some Daphnia will probably slow down with about four drops of copper sulfate.

d. How many drops of solution kill the organisms?

Some Daphnia will probably die with seven or more drops of copper sulfate.

e. From the results of this experiment, can you predict what effect copper sulfate in the water supply would have on other life forms? If no, why not? If yes, what would you predict?

No; the results of this experiment do not provide the data needed to make predictions about the effect of copper sulfate on other life forms. Only one kind of animal and no plants were tested, and the length of study was very short, so no generalizations can be substantiated.

Report format: Answer the questions above.

A C T I V I T Y 30

Food Additives

Purpose: To identify food additives and their functions

Time required: 2 hours outside class

Materials

ingredient labels on food products

Background information: Food manufacturers use hundreds of additives, including preservatives, nutritional supplements, flavoring agents, coloring agents, emulsifiers, stabilizers, and thickeners. Federal law requires that food labels list these and other ingredients.

Preservatives include salt, BHA (butylated hydroxyanisole), BHT (butylated hydroxytoluene), and EDTA (ethylene diamine tetraacetic acid). Nutritional supplements include elements (calcium, for example), minerals, and vitamins. Flavoring agents include spices and natural fruit flavors, as well as such artificial flavors as vanillin. Among the various dyes is beta carotene. Carrageenan—a stabilizer—is extracted from seaweed. Monoglycerides, diglycerides, and polysorbates are examples of emulsifiers. Pectin and gelatin are thickeners.

Name ______________________ Class ______________ Date ______________

Procedure:

1. Research the functions of the food additives in the categories named above and list several widely used additives in each category.

Preservatives inhibit the growth of microorganisms or prevent fats from spoiling: salt, BHA, BHT, EDTA. Nutritional supplements replace nutrients lost in processing: minerals, vitamins, elements (like calcium and iron). Flavorings enhance taste: spices, vanillin. Colorings improve appearance: beta carotene, beet juice. Stabilizers improve consistency and texture: carrageenan. Emulsifiers enable oil and water to mix: lecithin, monoglycerides, diglycerides. Thickeners improve texture: gelatin, pectin.

Name ______________________ Class ____________ Date ____________

2. At home or in a grocery store, look for the additives on your list on packages of the following: cereal, ice cream, hot dogs, a frozen dinner, and margarine.

3. Record your findings on Table A by putting a + sign in each appropriate box to indicate the presence of an additive.

Table A

	Cereal	Ice cream	Hot dogs	Frozen dinner	Margarine
Preservatives					
Supplements					
Flavorings					
Colorings					
Emulsifiers					
Stabilizers					
Thickeners					

Questions:

1. List some advantages and disadvantages of putting additives in food products. Do the benefits outweigh the drawbacks? Why?

Additives increase product shelf life; improve taste, texture, and appearance; and decrease food costs. Long-term effects of some additives are not known and may be undesirable. (Students may list other advantages and disadvantages. Personal choices determine whether benefits outweigh drawbacks or not.)

2. Will you change what you eat as a result of your research about additives?

(Various answers.)

3. Discuss the subject of food additives with your family and friends. re they aware of the different additives and their functions?

(Various answers.)

Report format: Answer the questions and turn in your chart.

ACTIVITY

Government and the Environment

31

Purpose: To identify an environmental issue of interest to you and to communicate your interest to an appropriate public official

Time required: 1 period; 3 to 4 hours outside class

Materials

paper and pen

Background information: Topics might include the quality of local drinking water, clear-cutting of state timber stands, offshore oil drilling, acid rain, protection of endangered species, the future use of public lands, and disposal of hazardous wastes.

Name ______________________ Class ____________ Date ____________

Procedure:

1. List four or more environmental issues that interest you.

(Various answers.)

2. Select one issue that particularly concerns you. Research the facts and various positions pertaining to it. State the main points of controversy.

(Various answers.)

Name ______________________ Class ____________ Date ____________

3. Write a letter expressing your views on the subject, being sure to provide sound reasons for your position. Use a positive tone and have practical suggestions.

4. Find out the proper person(s) to send the letter to, keeping in mind that several agencies may be responsible for some aspect of the issue. Public officials you might contact include the city manager, mayor, county manager, county commissioner or supervisor, state assembly member or senator, U.S. representative or senator, or governor. Find out the proper form of addressing the letter and the address to which you will send it. Keep a copy.

5. If you receive a response, describe the position stated. If you do not receive a response, send another copy of the letter or address it to someone else.

(Various descriptions.)

____

____

6. Participate in a class discussion summing up the experience and what you have learned about communicating with government officials about environmental issues.

Report format: Turn in a copy of your letter and any responses.